T0142469

Springer Theses

Recognizing Outstanding Ph.D. Research

Aims and Scope

The series "Springer Theses" brings together a selection of the very best Ph.D. theses from around the world and across the physical sciences. Nominated and endorsed by two recognized specialists, each published volume has been selected for its scientific excellence and the high impact of its contents for the pertinent field of research. For greater accessibility to non-specialists, the published versions include an extended introduction, as well as a foreword by the student's supervisor explaining the special relevance of the work for the field. As a whole, the series will provide a valuable resource both for newcomers to the research fields described, and for other scientists seeking detailed background information on special questions. Finally, it provides an accredited documentation of the valuable contributions made by today's younger generation of scientists.

Theses are accepted into the series by invited nomination only and must fulfill all of the following criteria

- They must be written in good English.
- The topic should fall within the confines of Chemistry, Physics, Earth Sciences, Engineering and related interdisciplinary fields such as Materials, Nanoscience, Chemical Engineering, Complex Systems and Biophysics.
- The work reported in the thesis must represent a significant scientific advance.
- If the thesis includes previously published material, permission to reproduce this must be gained from the respective copyright holder.
- They must have been examined and passed during the 12 months prior to nomination.
- Each thesis should include a foreword by the supervisor outlining the significance of its content.
- The theses should have a clearly defined structure including an introduction accessible to scientists not expert in that particular field.

More information about this series at http://www.springer.com/series/8790

Yangmei Li

Studies of Proton Driven Plasma Wakefield Acceleration

Doctoral Thesis accepted by
University of Manchester, Manchester, UK

Author
Dr. Yangmei Li
Advanced Interdisciplinary Technology Research Center
National Innovation Institute of Defense Technology
Beijing, China

Supervisor
Prof. Guoxing Xia
School of Physics and Astronomy
University of Manchester
Manchester, UK

ISSN 2190-5053 ISSN 2190-5061 (electronic)
Springer Theses
ISBN 978-3-030-50118-1 ISBN 978-3-030-50116-7 (eBook)
https://doi.org/10.1007/978-3-030-50116-7

This Springer imprint is published by the registered company Springer Nature Switzerland AG
The registered company address is: Gewerbestrasse 11, 6330 Cham, Switzerland

Supervisor's Foreword

Particle accelerators are indispensable tools in fundamental research and in applications widely used in industry and medicine. For particle physics, accelerators play the most important role in finding the basic building blocks of our Universe and unlocking the underlying physics principles describing their interactions. From Rutherford's first splitting of the atom to nowadays, accelerator science has undergone a glorious century. It has made the greatest contributions to the advancement of modern physics.

All of these operational accelerators are rooted in conventional Radio Frequency (RF) technology. However, this technology has reached its limit today mainly due to very large sizes and costs of these facilities, especially for those in high energy particle physics research. For instance, the Large Hadron Collider (LHC) at the European Organization for Nuclear Research (CERN), a circular machine with a circumference of 27 km and the construction costs of over $4 billion, took about a decade to construct. The recently proposed Future Circular Collider (FCC) would be the most powerful particle smasher ever built. It would be a circular machine with 100 km circumference and construction costs of up to about $26 billion.

New acceleration methods, with the aim of achieving a more efficient and more affordable solution for particle acceleration, are definitely needed. Fortunately, two pioneering scientists, Tajima and Dawson were the first to propose utilizing the large collective fields in plasmas to propel particles to high energies in late 1979. Since then, these plasma-based accelerators have developed and achieved tremendous progress in laboratories worldwide. Nowadays, the wakefield acceleration using short-pulse and high-power lasers, so-called Laser Wakefield Acceleration (LWFA) can routinely achieve $\sim$GeV electron beams in centimetre-long plasma cells, while the relativistic electron or positron beam, if serving as the drive beam in a so-called Plasma Wakefield Accelerator (PWFA), can excite plasma wakefield and efficiently accelerate a trailing beam to high energy in a plasma cell of tens of centimetres or a metre in length. The resultant wakefield amplitudes for the above two schemes are about 10–100 GV/m, which are 2 or 3 orders of magnitude higher than the accelerating fields in conventional RF-based accelerators (usually less than 100 MV/m).

Due to the limited energy of current' laser pulses and electron beams ($\sim$0.1 kJ per pulse), the future TeV collider design based on either LWFA or PWFA will have to combine multiple stages to ramp up the energy step by step. This will complicate the overall design and pose stringent requirements on synchronization and alignment of the system. On the other hand, the proton bunches, especially those extracted from the CERN accelerator complex, such as Super Proton Synchrotron (SPS) or the LHC can provide proton bunches with energies from 10 kJ to a maximum 140 kJ. If such a proton bunch is employed to drive the plasma wakefield, it would have great potential to accelerate an electron beam to the energy frontier, i.e. TeV in a single acceleration stage. Fortunately, this has been successfully demonstrated by simulation. The recent AWAKE (Advanced Wakefield Experiment) experiment at CERN also observed up to 2 GeV electron acceleration using the SPS proton bunch as drive beam.

Yangmei's work is related to this exciting research area on Proton-Driven Plasma Wakefield Acceleration (PD-PWFA). She joined our research group in September 2015 as the President Doctoral Scholarship (PDS) recipient. From 2015 to 2019, Yangmei has done excellent research in understanding the underlying physics of proton–plasma interactions. In her thesis, she used the theory of beam–plasma interactions and particle-in-cell simulations, to demonstrate that high-energy, high-quality electrons and positrons can be achieved when a high-energy single short proton bunch or a proton bunch train is used as drive beam. In her studies, she investigated the physics in hollow plasma channels, which have obvious advantages compared to uniform plasma. The resultant longitudinal accelerating wakefield and negligible transverse wakefield in a hollow plasma are ideal for the acceleration of both electrons and positrons. The studies show that the hollow plasma configuration greatly helps the preservation of the beam energy spread and final emittance, which are the figures of merit for future energy frontier colliders.

In addition, Yangmei also explored the transverse instabilities induced by misaligned beams in hollow plasma and the enhancement of wakefield amplitude driven by a self-modulated proton bunch with a tapered plasma density. This scheme (a tapered plasma density) is currently used in the AWAKE experiment at CERN. The results she achieved in this thesis have major potential to impact the future of the AWAKE project and the next generation of linear colliders, and also, in the long-term, may allow compact accelerators to be used in industrial and medical facilities.

Manchester, UK
July 2020

Prof. Guoxing Xia

Abstract

Proton driven plasma wakefield acceleration features both ultra-high accelerating gradients thanks to plasmas and long acceleration distances due to high-energy contents of protons. Hence, it has great potential of powering leptons to TeV-scale energies in a single plasma stage. Nonetheless, the plasma response is different from that driven by the electron bunches or lasers as protons suck in the plasma electrons rather than repel them. It leads to radially nonlinear and time-varying focusing at the accelerating region, which is detrimental to the witness beam quality.

In this thesis, by theory and particle-in-cell simulations, we demonstrate the viability of generation of high-energy, high-quality electrons or positrons driven by a single short proton bunch or a proton bunch train, which paves the way for the development of future energy frontier lepton colliders. The hollow plasma channel has been involved in our studies, which plays a vital role in creating a beneficial accelerating structure. It eliminates the transverse plasma fields and forms radially uniform accelerating fields. The former benefits preservation of the beam emittance. The latter enables minimization of the energy spread by longitudinal control of the beam.

The relating issues have been investigated, for example, the transverse effects induced by the beam-hollow misalignment, and the wakefield excitation driven by a practically long proton bunch in the uniform plasma, where the concept of seeded self-modulation has been employed into the AWAKE experiment.

Publications related to this thesis

1. **AWAKE Collaboration**, "Acceleration of electrons in the plasma wakefield of a proton bunch", *Nature*, vol. 561, 363, 2018.

2. M. Turner et al. (**AWAKE Collaboration**), "Experimental observation of plasma wakefield growth driven by the seeded self-modulation of a proton bunch", *Phys. Rev. Lett.*, vol. 122, 054801, 2019.

3. E. Adli et al. (**AWAKE Collaboration**), "Experimental observation of proton bunch modulation in a plasma at varying plasma densities", *Phys. Rev. Lett.*, vol. 122, 054802, 2019.

4. **Y. Li**, G. Xia, K. V. Lotov, A. P. Sosedkin, Y. Zhao, "High-quality positrons from a multi-proton bunch driven hollow plasma wakefield accelerator", *Plas. Phys. Control. Fusion*, vol. 61, 025012, 2019.

5. **Y. Li**, G. Xia, K. V. Lotov, A. P. Sosedkin, Y. Zhao, S. J. Gessner, "Amplitude enhancement of the self-modulated plasma wakefields", *J. Phys. Conf. Ser.*, vol. 1067, 042009, 2018.

6. **Y. Li**, G. Xia, K. V. Lotov, A. P. Sosedkin, K. Hanahoe, O. Mete-Apsimon, "Multi-proton bunch driven hollow plasma wakefield acceleration in the non-linear regime", *Phys. Plasmas*, vol. 24, 103114, 2017.

7. **Y. Li**, G. Xia, K. V. Lotov, A. P. Sosedkin, K. Hanahoe, O. Mete-Apsimon, "High-quality electron generation in a proton-drive hollow plasma wakefield accelerator", *Phys. Rev. Accel. Beams*, vol. 20, 101301, 2017.

8. **Y. Li**, G. Xia, Y. Zhao, "Assessment of transverse instabilities in proton driven hollow plasma wakefield acceleration", *Proc. IPAC18*, (Vancouver, Canada), TUPML022, 2018.

9. **Y. Li**, G. Xia, K. Hanahoe, T. Pacey, O. Mete, "Proton-driven electron acceleration in a hollow plasma", *Proc. IPAC16*, (Busan, Korea), TUPMY025, 2016.

10. K. Hanahoe, G. Xia, M. Islam, **Y. Li**, O. Mete-Apsimon, B. Hidding, J. Smith, "Simulation study of a passive plasma beam dump using varying plasma density", *Phys. Plasmas*, vol. 24, 023120, 2017.

11. G. Xia, Y. Nie, O. Mete, K. Hanahoe, M. Dover, M. Wigram, J. Wright, J. Zhang, J. Smith, T. Pacey, **Y. Li**, Y. Wei, C. Welsch, "Plasma wakefield acceleration at CLARA facility in Daresbury Laboratory", *Nucl. Instrum. Meth. Phys. Res. Sect. A*, vol. 829, pp. 43–49, 2016.

12. A. Caldwell et al. (**AWAKE Collaboration**), "Path to AWAKE: Evolution of the concept", *Nucl. Instrum. Meth. Phys. Res. Sect. A*, vol. 829, pp. 3–16, 2016.

13. E. Gschwendtner et al. (**AWAKE Collaboration**), "AWAKE, The Advanced Proton Driven Plasma Wake field Acceleration Experiment at CERN", *Nucl. Instrum. Meth. Phys. Res. Sect. A*,vol. 829, pp. 76–82, 2016.

14. J. Resta-López, A. Alexandrova, V. Rodin, Y. Wei, C. P. Welsch, **Y. Li**, G. Xia, Y. Zhao, "Study of ultra-high gradient acceleration in carbon nanotube arrays", *Proc. IPAC18*, (Vancouver, Canada), TUXGBE2, 2018.

Acknowledgments

I am trying not to be sentimental but fail when recalling my three-and-a-half-year life and study in the UK. I feel myself so lucky because of all the excellent people in my life and a great deal of support I have received from them. I would like to express my great gratitude to them on the occasion of finishing my thesis.

I would like to first thank my super nice supervisor Dr. Guoxing Xia. I would never have possibly completed my study without all his patient guidance and selfless support. I appreciate the huge amount of time he makes for me, much of which is actually squeezed out from his pretty occupied work and from his holiday. He is more like a thoughtful and caring friend than a "boss", which makes me free to talk about my anxieties and worries with him. He always tries his best to help. I am grateful for all the opportunities he created for me to discuss with experts in the cutting-edge fields, develop my communication skills and broaden my eyes. More thanks should be given to him for bringing me on to the marvelous AWAKE collaboration, so that I have the chance to work with distinguished theoretical and experimental scientists on the world's first proton -driven plasma wakefield acceleration experiment, and to witness with great excitement the first acceleration of electrons right before I graduate.

I would like to specially thank my important collaborators and also the developers of the particle-in-cell code LCODE I mainly use: Prof. Konstantin Lotov and Dr. Alexander Sosedkin from Budker Institute of Nuclear Physics. I have had considerable discussions with them on my simulation results and got plenty of prompt technical support in utilising many advanced features of LOCDE and running it on the clusters. I deeply admire Konstantin's rich knowledge of physics and his capability of quickly capturing the pivotal points of the physical issues. His comments and suggestions on my work have motivated me to explore many further studies, which form part of this thesis. Apart from helping revise my manuscripts, he also offered me many tips about writing good academic papers. Alexander although works in the physics department, is definitely an outstanding programmer. He seems get to know all the answers to my coding or Unix issues in no time, which I might need to spend a day or even a week on.

Many thanks to Kieran who has always been very helpful and patient. He used to explain the physics to me over and over until I fully understood. Even when he was busy writing his thesis, he still spared much time helping me get started on using EPOCH and Vsim. Despite that he has graduated from the university, he is still highly responsive with a lot of details to my questions. Special thanks to Oznur, the postdoc in our group for her many constructive suggestions on my work and for her much time devoted into promoting my papers and thesis. More thanks go to my peer colleagues from the University of Manchester and Cockcroft Institute: Yelong, Tom, Bill, Toby, Agnese, Taaj. We have developed good friendships while working on group projects in the Cockcroft Institute accelerator school, playing room escape games, climbing stones, enjoying food and drinks, and chatting on the way from Manchester to Daresbury tons of times, etc.

I want to thank all the academic staff in the accelerator physics group, especially the head of group Prof. Roger Jones who has been very interested in my work and concerned about my progress. Not long ago, he invested many efforts into the document preparation in order to nominate me as the research postgraduate of year.

Special thanks to the whole AWAKE collaboration for the numerous efforts they make, the great perseverance and enthusiasm they show while doing successive day and night experimental shifts for months, the creative ideas they come up with in dealing with massive issues. The huge success in Run 1 cannot be possibly obtained without all of their contributions. I feel privileged to be one of this vigorous team. I have profited a lot from their scientific thinking, deep physics insights, excellent presentation skills and nice personalities. I have precious and fun memories with them in Switzerland, Novosibirsk, Elba island and Manchester. I will cherish the friendships with many of them forever: Spencer, Alexey, Barney, Anna-maria, Fabian, Mathias, Alexander, Petr, Vladimir, Aravinda, Gabriel, Florence, Michele.

I also feel so grateful and lucky being accompanied by many good Chinese friends while studying abroad: Mengying, Qin, Fan, Yuan, Dongjiao, Pengzhan, Chunhu, Yang, Wenjun, Sen, Chunren, Rui, Emiao at Manchester, Huizhe, Wenfeng, Rongjuan, Xiangwen, Ningning at Druham, Mengrong, Qian at Bath, Xue at Leeds. I appreciate all the emotional support they gave me. I have priceless memories with them in swimming, park running, hiking, cooking Chinese food, playing board games, travelling across the UK. My overseas life couldn't be so much colourful without these great pals.

The acknowledgements cannot be written without mentioning the Reslife team, which I have worked with for one and a half years in promoting students' living experience in the university accommodation. We endeavour together to support the students. More importantly, we are each other's backup.

Many thanks to Prof. Zhengming Sheng and Prof. Philippa Browning for accepting to be my thesis examiners. I appreciate a lot their time and all the comments from them to promote my thesis.

My sincere gratitude goes to my beloved family who has always been there supporting me, understanding me, encouraging me and caring me. They are my driving force in pursuing further studies and facing difficulties bravely.

I would like to thank my fiance Dr. Shaoyan Sun. Thanks for his countless encouragement, understanding and patience. His integrity and high sense of responsibility have always been motivating me. In the past 10 years, especially in the 6-year long-distance relationship, we have never stopped supporting each other. Now I am joyfully looking forward to our reunion and the new life ahead of us.

Last but not least, I greatly acknowledge the financial support from the President's Doctoral Scholar Award, the Cockcroft Core Grant and STFC, also the massive computing resources provided by the clusters of SCARF, CSF, N8HPC, and JURECA.

Contents

Symbols

c	Speed of light in vacuum (299 792 458, m s^{-1})
e	Elementary charge (1.602×10^{-19}, C)
m_e	Electron mass (9.109×10^{-31}, kg)
$m_e c^2$	Electron mass-energy (0.511, MeV)
m_p	Proton mass (1.673×10^{-27}, kg)
m_p/m_e	Proton-electron mass ratio (1836.153, –)
$m_p c^2$	Proton mass-energy (0.938, GeV)
μ_0	Vacuum permeability ($4\pi \times 10^{-7}$, H m^{-1})
ϵ_0	Vacuum permittivity (8.854×10^{-12}, F m^{-1})
e	Exponential constant (2.718)

Notations

t	Propagation time (s)
z	Propagation distance (m)
ξ	Co-moving coordinate (m)
γ	Relativistic factor (–)
β	Ratio of relativistic v to c (–)
q	Electric charge (C)
Q	Total electric charge (C)
ρ	Charge density (C m^{-3})
I	Electric current (A)
J	Current density (A m^{-2})
E	Electric field (V m^{-1})
ϕ	Electric scalar potential (V)
B	Magnetic field (T)
A	Magnetic vector potential (T m)
n_e	Electron number density (m^{-3})

n_0	Background plasma electron density (m^{-3})
n_i	Background plasma ion density (m^{-3})
n_1	Perturbed plasma electron density (m^{-3})
ω_p	Plasma electron frequency (rad s^{-1})
λ_p	Plasma wavelength (m)
k_p	Plasma wavenumber (rad m^{-1})
c/ω_p	Collisionless plasma skin depth (m)
$m_e c\omega_p/e$	Plasma wave-breaking field (V m^{-1})
r_c	Hollow plamsa channel radius (m)
N	Beam population (–)
σ_r	RMS beam radius (m)
σ_z	RMS beam length (m)
ϵ	Geometrical emittance (m rad)
ϵ_n	Normalized emittance (m rad)
W	Energy (eV)
$\delta W/W$	Energy spread (–)
L_q	Quadrupole magnet period (m)
S	Magnetic field gradient (T m^{-1})
$\mathcal{L}$	Luminosity (m^{-2} s^{-1})

Chapter 1
Introduction

1.1 Conventional Accelerators

High energy physics is devoted to exploring the fundamental structure of the physical world and studying the basic laws that govern our universe. By means of particle collisions, we have been able to investigate initially the structure of an atom, then discover the nucleus and subsequently a variety of new particles and their interactions. For example, with the world's highest-energy (7 TeV) particle collider, the Large Hadron Collider (LHC, Fig. 1.1a) built at CERN (the European Organization for Nuclear Research), the Higgs Boson was discovered in 2012 [1, 2]. The larger energies the colliding particles possess, the more new particles can be created and physics can be studied. In addition, as leptons are point-like and fundamental objects, a significantly cleaner collision environment is achievable in lepton colliders in comparison with hadron colliders and hence higher precision for physics measurements. As a result, it is widely held that the next energy frontier colliders should collide electrons and positrons in the Teraelectronvolt (TeV, 10^{12} eV) scale.

Conventional radio-frequency (RF) accelerators have been the main tools to accelerate particles to high energies with the RF fields sustained in the metallic cavities. However, limited by the breakdown of the metallic surface walls [4], the maximum accelerating gradient hasn't surpassed 100 MeV/m. Therefore, in order to obtain particle energies of tens of GeV or even TeV, the dimensions of accelerators are supposed to be at least hundreds to tens of thousands of meters long. For instance, the accelerating facilities at the Stanford Linear Accelerator Center (SLAC) in USA are as long as 3 km to accelerate electrons or positrons to 50 GeV. The average accelerating gradient is even less than 20 MeV/m. Circular machines like synchrotrons are adopted to compact the accelerators, where the magnets are used to keep the particles in a circular orbit so that the particle energies can be boosted many turns in the ring. For example, the LHC can produce protons up to 7 TeV. However, circular machines are not feasible for producing TeV light particles like electrons and positrons due to enormous energy loss from synchrotron radiation, which is caused by the magnetic fields bending the charged particles. The radiation loss follows the relation [5]

Y. Li, *Studies of Proton Driven Plasma Wakefield Acceleration*, Springer Theses,
https://doi.org/10.1007/978-3-030-50116-7_1

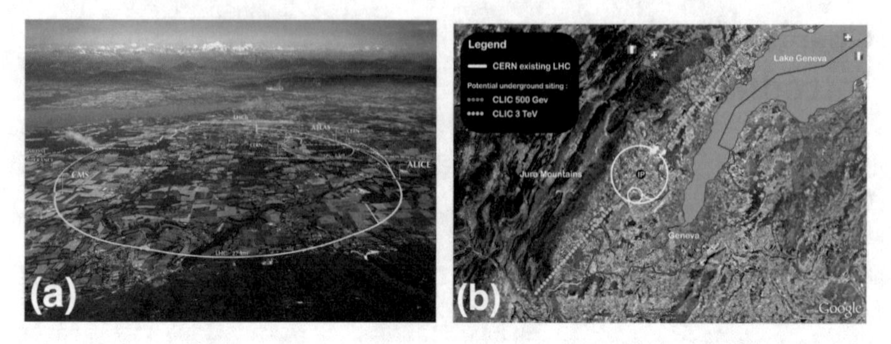

Fig. 1.1 Aerial views of the LHC (**a**) and a potential location for the CLIC accelerator complex on a map (**b**). Images reproduced from the website of CERN and from the design report for CLIC [3] respectively

$$W_{loss} \propto \frac{\gamma^4}{\rho}, \tag{1.1}$$

where $\gamma = W/(mc^2)$ is the particle's relativistic factor, W and m are the particle energy and mass respectively, and c is the speed of light. Due to strong γ dependence, the electrons will radiate away 1.13×10^{13} times more energy than the protons if they are accelerated to the same energy. Hence, the future TeV lepton colliders must rely on linear accelerators. Here it is worth pointing out that the enormous circumference of the accelerating ring of the LHC (27 km, Fig. 1.1a) is not because of the limit of the accelerating gradient. It comes from the strength of dipole magnets installed limited to $B = 8.33$ T, which restricts the bending radius ρ given a nominal proton energy W due to the relation

$$W \propto B\rho. \tag{1.2}$$

Two energy frontier linear collider candidates have been proposed to accelerate leptons: the International Liner Collider (ILC) and the Compact Linear Collider (CLIC, Fig. 1.1b). The ILC is designed to boost the electron and positron energy to 500 GeV for collisions and later possibly upgraded to 1 TeV. With a planned accelerating gradient of 31.5 MeV/m on average utilising super-conducting RF cavities, the total footprint is expected to be between 30 km and 50 km [6]. The CLIC exploits a novel two-beam acceleration technique where a low energy, high peak current drive bunch transfers its power to the RF cavities where the short main beam with low current is accelerated [3]. It is expected to promote the field gradient in the accelerating structure to as high as 100 MeV/m and generate centre-of-mass energies up to 3 TeV. The accelerator would be between 11 km and 50 km long. While the sizes and costs are kept within reach, they are still more than ten times larger than for the existing SLAC accelerating facilities. It is not surprising that such projects are funded under collaborative studies around the world. The sizes and construction costs of such huge linear accelerators would be prohibitive for every single country. Also it would be challenging to find suitable locations. All these difficulties urge scientists to seek alternative acceleration schemes which can considerably shrink the sizes of the machines and reduce the expenses.

1.2 Plasma-Based Accelerators

Plasma is an electrically neutral medium consisting of ions and free electrons which exhibit collective effects [7]. As an already ionized gaseous substance, it can sustain large electric fields. The maximum field, so-called wave-breaking field E_{wb} when the plasma electron wave breaks under an intense perturbation follows the relation

$$E_{\mathrm{wb}} = m_{\mathrm{e}} c \omega_{\mathrm{p}} / e = 96 \sqrt{n_{\mathrm{e}}\,[\mathrm{cm}^{-3}]}\,(\mathrm{V/m}), \tag{1.3}$$

where $\omega_{\mathrm{p}} = \sqrt{e^2 n_{\mathrm{e}} / \epsilon_0 m_{\mathrm{e}}}$ is plasma electron wave frequency, n_{e} is the plasma electron density, m_{e} is the electron mass, e is the elementary charge, c is the speed of light, ϵ_0 is the vacuum permittivity. Given a plasma electron density of $10^{18}\,\mathrm{cm}^{-3}$, the cold plasma wave-breaking field reaches up to 100 GV/m, which is three orders of magnitude higher than the electric breakdown threshold of conventional accelerators. This implies that plasma is a promising medium for ultra-high gradient acceleration of charged particles.

1.2.1 Laser Driven Accelerators

In 1979, Tajima and Dawson [8] firstly put forward the concept of plasma-based particle accelerators [9]. An intense laser pulse shines the underdense plasma and excites plasma electron oscillations via its ponderomotive force (now known as laser wakefield acceleration (LWFA) [23]). The ponderomotive force results from rapidly oscillating, non-uniform laser fields and expels the plasma electrons away from the axis while the ions are considered immobile due to large inertia in this short time scale. The electric fields formed behind the laser pulse (i.e., "wake" fields) due to the charge separation of plasma can accelerate trapped electrons to relativistic energies. The most effective way requires the laser pulse length to be half the plasma wavelength ($\lambda_{\mathrm{p}} = 2\pi c / \omega_{\mathrm{p}}$), which was challenging for the existing glass lasers. Hence, an alternative approach was proposed: plasma beat-wave accelerator (PBWA) [24]. The principle is to excite the plasma wave by two co-propagating laser pulses with a slight frequency difference $\Delta\omega$. Resonant excitation occurs when the frequency difference is approaching the plasma frequency. High plasma fields are therefore generated. Another way to address the long laser pulse issue is the self-modulated LWFA (SM-LWFA) [10, 11]. In this scheme, local oscillations of the plasma electrons modulate the long laser pulse ($c\tau \gg \lambda_{\mathrm{p}}$) into a pulse envelope similar to that in the PBWA scheme during the laser-plasma interaction. The modulation is resonant throughout the driving pulse and a positive feedback is developed to further increase the amplitude of the plasma wave.

Following the development of laser technologies, particularly the emergence of chirped-pulse amplification (CPA) [12], single, short (<1 ps), ultraintense ($>10^{18}$ W/cm^2) laser pulses [13, 14] are available for driving strong plasma wakefields. A "bubble" regime [15, 16] is then formed where the ponderomotive force of the laser pulse is strong enough to expel all the plasma electrons from the axial region, leaving an ion volume without electrons. The "bubble" forms an ideal field structure for electron acceleration in three aspects. First of all, the longitudinal electric fields at the witness electron bunch are independent of the radial positions, thus favourable to reduce the energy spread. Secondly, the transverse electric fields from the background uniform ions increase linearly with the radius, which form an ideal focusing lens. It benefits the preservation of the electron transverse emittance. Finally, the accelerated electrons and the driving laser pulse are spatially separated from each other, which is advantageous for stable laser propagation and conserving the electron beam quality as well [11]. The first experimental results in or close to the "bubble" regime were reported in 2004 [17–19] and near-monoenergetic electron bunches were obtained for the first time. This is considered as a milestone in the development of LWFA. After that, the LWFA has advanced extensively [20, 21]. The generated electron bunches nowadays are comparable with that needed in the synchrotron light sources regardless of some limitations such as lower repetition rates, larger energy widths, larger shot-to-shot jitter, etc. The latest energy record of 4.2 GeV was created by Lawrence Berkeley National Laboratory (LBNL) in 2014 [22]. In the experiment, a 0.3 PW Ti: sapphire-based BELLA laser was employed to drive a capillary-discharge-guided plasma with a density of 7×10^{17} cm^{-3} over 9 cm. The accelerating gradient demonstrated was over 46 GeV/m. In the near future, they aim at obtaining a higher energy output -10 GeV with a new technique of plasma-channel shaping.

Despite huge advantages, such as ultra-high accelerating gradients, hence considerably short accelerating distances and linear focusing fields for witness bunches that the LWFA brings, it is subject to three factors which limit the maximum acceleration length and therefore the achievable energy gain of charged particles [23]. The first and the most serious one is natural diffraction of the laser pulse during propagation as it is significantly focused to obtain high intensity before injected into the plasma. It limits the interaction distance to the Rayleigh length. In addition, depletion of the laser pulse due to continuous energy exchange to plasma waves sets a limit to the depletion length. Exploitation of external guiding [11, 18–20, 25] and higher energy laser pulses can relax these two limitations to a certain extent. Still, when the charged bunch is accelerated to the (ultra-) relativistic energy, its velocity ($\sim c$) will surpass the laser group velocity (less than c). As a result, the bunch overruns the plasma wave and gets into the decelerating region. To further promote the energy gain, multi-staging acceleration [26, 27] has been proposed, whereas it demands precise alignment and synchronisation of each stage and between the drivers and the accelerated beam, thus is technically challenging in experiment [28–30].

1.2.2 Electron Driven Accelerators

In 1985, Chen et al. [31] first suggested that the driving laser could be replaced by the electron beam. They introduced two bunched electron beams into a cold plasma and demonstrated the successful acceleration of driven electrons from $\gamma_0 m_e c^2$ to $3\gamma_0 m_e c^2$ via simulations. Unlike LWFA, here it is the space charge force from the charged driving bunch that expels free plasma electrons and causes plasma electron oscillations along the axis. A trailing bunch is accelerated by the electric fields resulting from plasma charge separation. This is the prototype of beam driven plasma wakefield acceleration (PWFA). PWFA outperforms LWFA in the following ways. First of all, the relativistic driving bunch travels at the speed of light, it therefore does not dephase with the relativistic witness bunch. Secondly, it is strongly focused by the transverse plasma wakefields thus hardly gets diffracted. Furthermore, the energy content of a beam driver is determined by the particle population and the energy per particle. As a result, it is easier to promote the driving energies and a significantly longer depletion length is achievable in experiment.

In the linear regime of PWFA ($n_b \ll n_e$, $E/E_{wb} \ll 1$), the longitudinal plasma electric field E (in the position of $z - ct$) behind the driving bunch is sinusoidal with the plasma frequency, and the peak acceleration amplitude when $k_p\sigma_z \cong \sqrt{2}$, $k_p\sigma_r \ll 1$ can be written as [32]:

$$E_{max} = 240\,\mathrm{MV/m}\left(\frac{N}{4\times10^{10}}\right)\left(\frac{0.6}{\sigma_z\,(\mathrm{mm})}\right)^2. \tag{1.4}$$

Here n_b is the peak driving bunch density, n_e is the unperturbed plasma electron density, $k_p = \omega_p/c$ is the plasma wave number, z is the beam travelling distance, σ_z and σ_r are the RMS (root mean square) bunch length and radius respectively. It follows that the maximum acceleration gradient is proportional to the number of driving particles N and inversely proportional to the square of σ_z. This applies equally to electron and positron drivers. The corresponding perturbed wakefields also behave the same except for a wake phase difference of π.

In the nonlinear regime ($n_b > n_e$), the wakefield excited by a dense electron driver resembles the "bubble" or "blow-out" structure in LWFA. In fact, the analogous investigation actually starts from the electron driven case [33, 34]. The high density electron driver repels all the plasma electrons from the bunch volume and forms an electron-free "bubble" filled with uniform ions. The transverse focusing is linearly increasing along the radius and is constant along the witness bunch. The longitudinal accelerating field is uniform along the radius. The combination of aberration-free, strong focusing and a large accelerating gradient makes this acceleration scheme ideal for the witness bunch in terms of high energy gain and well-preserved beam emittance in a long distance.

It is worth pointing out that the positively charged drivers like positrons behave differently in the nonlinear regime [32]. Instead of being expelled, the plasma electrons are pulled towards positrons and then stream through them. The resulting focusing is nonlinear radially and varies along the witness bunch. In addition, the accelerating gradient is less than in the electron driven case under the same condition because the streaming plasma electrons do not return as an ensemble to the propagation axis.

A lot of work concerning electron driven acceleration has been done in the aspects of theory, simulation and experiment since the proposal of PWFA [33–43]. In 2007, an inspiring experiment completed at the SLAC reported an energy doubling of the rear fraction of a 42 GeV driving electron bunch over less than 1 m long plasma [42]. The peak accelerating gradient reached up to ~ 52 GeV/m. Although the final energy spread was almost 100%, this work still profoundly revealed the feasibility of plasma accelerators towards high energy physics applications. As this single bunch scheme only allowed a small number of tail particles to be accelerated, it has been proposed to introduce a distinct bunch containing a substantial charge to further extract energy from the high gradient wakefields [43]. Figure 1.2 illustrates the concept where a trailing electron bunch is placed at the maximum accelerating region. The experiment demonstrated a high energy transfer efficiency of over 30% from the wake to the bunch as well as a low energy spread of 2%.

On the contrary, the progress on positron acceleration is slow [32, 44–47]. The first proof-of-principle experiment was conducted in 2003 where a single positron bunch was employed to drive the underdense plasma. An accelerating gradient of 56 MeV/m was measured at the tail of the positron bunch [45]. Then it took over a decade before the successful acceleration of positrons by 5 GeV over 1.3 m plasma was announced [44]. The most inspiring part of this work is that, about 10^9 positrons at the rear of the bunch out of the entire 1.4×10^{10} were accelerated with an energy spread as low as 1.8%, because the loading of high charge positron bunch flattens the accelerating field (Fig. 1.3). Overall, the success of accelerating electrons or positrons with both a high accelerating gradient and a high energy efficiency advances the development of compact and affordable plasma-based accelerators. Meanwhile it proves the PWFA concept to be an attractive scheme adopted in the plasma "afterburner" [48], where plasma is used to double the particle energy coming from the linear colliders.

1.2.3 Proton Driven Accelerators

One basic limitation for beam driven plasma wakefield acceleration is the transformer ratio. It is defined as the ratio of the energy gain of the witness bunch to the energy loss of the driver or the maximum accelerating field behind the driver to the maximum decelerating field acting on the driver. For longitudinal symmetric bunches, it cannot exceed 2 [35]. This indicates that the obtainable energy gain of the witness bunch in one acceleration stage cannot significantly exceed the energy of drive particles. The upper limit can be increased by employing non-symmetric bunches [49] or linearly ramped, equidistant driving bunch train [50], which is nonetheless either difficult or only works for flat-top beams.

Combining multiple acceleration stages can, in principle, overcome the transformer ratio limit, but electron-driven PWFA requires tens to hundreds of stages for

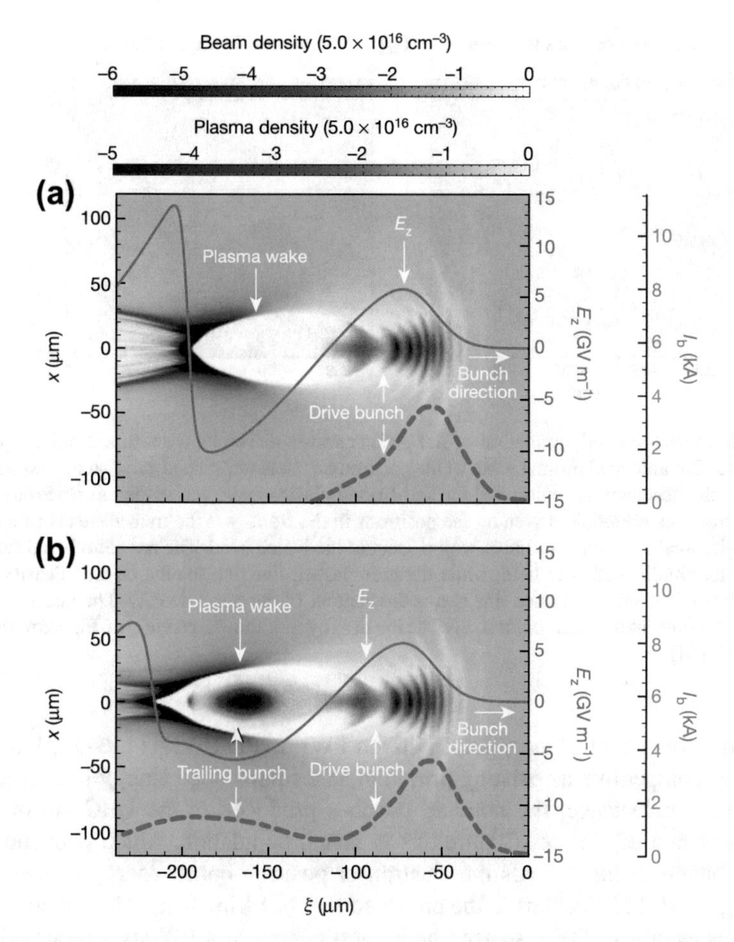

Fig. 1.2 Particle-in-cell simulation of the plasma waves driven by an electron beam without (**a**) and with (**b**) a trailing beam loaded. x is the transverse coordinate, ξ is the longitudinal position coordinate with respect to the driving bunch head. The red solid line denotes the on-axis longitudinal electric field, while the blue dashed line denotes the current of the beam. The blue and red color scales represent the plasma electron density and the beam density respectively. The ion density (not shown) is uniform. Figure reproduced from Ref. [43]

energy frontier collider applications, which is technically challenging as mentioned in Sect. 1.2.1. The need for multiple stages inherently arises from the limited energy contents of available drivers [51]. Laser driven wakefield accelerators (LWFA) are hindered by the same disadvantage. In spite of rapid growth of the laser peak power [52], single stage acceleration energy hasn't advanced that much in the last decade, going up from 1 GeV reached with 40 TW laser in 2006 [21] to 4.2 GeV record held from 2014 on [22]. Conservative designs of LWFA-based colliders thus also rely on staging [53–55].

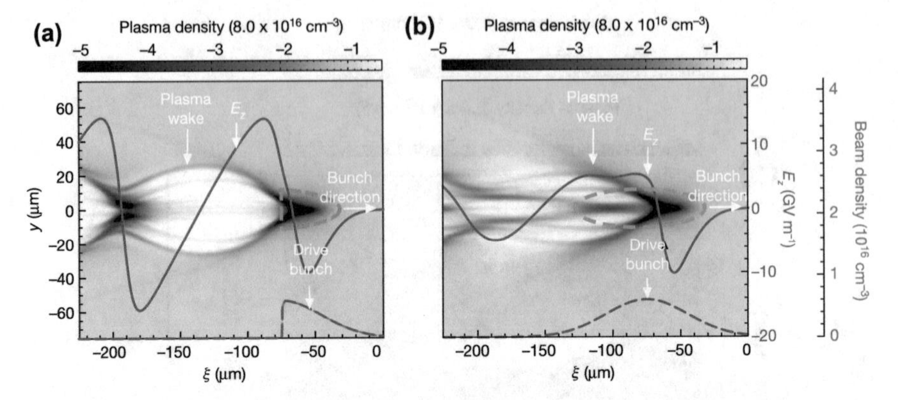

Fig. 1.3 Particle-in-cell simulation of the plasma waves driven by a short and intense positron bunch: (**a**) the unloaded plasma wave as the positron bunch is terminated early so that no positrons reside in the accelerating region; (**b**) the self-loaded plasma wave where the tail positrons extract energy from the wakefield driven by the positrons in the front. y is the transverse coordinate, ξ is the longitudinal position coordinate with respect to the bunch head. The red solid line denotes the on-axis longitudinal electric field, while the grey dashed line denotes the on-axis density profile of the beam. The orange dashed line marks the contour of the beam density. The plasma electron density is represented in blue color scale. The ion density (not shown) is uniform. Figure reproduced from Ref. [44]

Under this circumstance, proton driven PWFA (PD-PWFA) [56–59] looks particularly competitive as existing proton bunches have huge energy and large populations. For instance, the nominal bunches produced in the LHC are of 7 TeV in energy and of 1.15×10^{11} protons in bunch population, which corresponds to 129 kJ/bunch. It far exceeds the electron or positron bunch energy created at the SLAC [42] (0.12 kJ/bunch) or the proposed ILC (0.8 kJ/bunch). The energy of laser pulses is even less. For instance, the highest energy of a PW laser proposed to be used for wakefield acceleration is about 40 J [60]. Owing to available high energy contents, proton bunches become very promising candidates in the near future, which are capable of bringing witness particles to the energy frontier in a single plasma stage.

PD-PWFA was proposed by Caldwell et al. in 2009 for the first time [56]. Simulations prove that the electrons can be accelerated over hundreds of meters long plasma to TeV-level with an average gradient of 1.38 GeV/m. Detailed numerical analyses can be found in Ref. [57]. It should be noted that the plasma wave generates strong focusing at the tail of the driving bunch and the witness bunch, but very weak or no focusing at the head of the driving bunch. As a result, for the long distance propagation, quadrupoles are necessary to prevent the driver head from flying apart driven by the initial angular divergence.

1.2.4 *Self-modulated Long Proton Bunch Driven Accelerators*

The aforementioned PD-PWFA scheme assumes the proton bunches as short as hundreds of micrometers. However, currently existing high-energy proton bunches in synchrotrons are tens of cm long, which is much longer than the usual submillimeter plasma wavelength. For example, to get a wave breaking field in GeV/m, the plasma density needs to be at least 10^{14} cm^{-3}, which corresponds to a plasma wavelength of less than 3 mm. Hence, presently existing long proton bunches hardly excite strong plasma wakefields directly. Longitudinal compression of the bunch length to the plasma wavelength by traditional methods is conceivable [61, 62], but it will be too costly and technically challenging to implement, given that the required compression factor is up to several orders of magnitude. It therefore demands plenty of RF power to firstly generate the necessary energy chirp along the long proton beam and then huge dipoles for path modulation [61].

Fortunately, the seeded self-modulation (SSM) [62–65] offers a new way to cope with this issue. The SSM is the seeded axisymmetric mode of the transverse two-stream instability [66, 67] that develops in the beam-plasma system. The transverse wakefield of the beam leads to rippling of the beam itself, which further amplifies the plasma waves. Because of this positive feedback, the long proton bunch transforms into a train of equidistant micro-bunches that follow the plasma wavelength [68–70]. With proper seeding, the axisymmetric mode develops quickly and suppresses destructible non-axisymmetric modes like the hosing instability [71–73]. As a consequence, the bunched beam can resonantly drive strong plasma waves. The AWAKE (Advanced Wakefield) experiment at CERN [51, 74–76] relies on this concept and is the world's first proof-of-principle experiment of long proton bunch driven PWFA. So far it has successfully observed the seeded self-modulation of the long proton bunch [77, 78] and the acceleration of the injected electrons to 2 GeV in a 10 m long rubidium plasma [79] in the first run of experiments. The second run has been planned and will be launched in the year 2021 after two years' shut down of the LHC. Run 2 is dedicated to achieving a higher accelerating gradient and accelerating the electron beam over a longer distance and with high quality. In addition, several related experiments with long or pre-bunched electron beams [80–87] are proposed or conducted to enhance the understandings of the SSM and resonantly driven wakefields.

1.3 Motivations and Outline

1.3.1 *Motivations*

Plasma wakefield acceleration has been developing for four decades with the most attention fallen on the laser or electron driven cases. LWFA has been shown to be able to generate electron bunches which are comparable to that needed in the synchrotron light sources. The latest energy record is 4.2 GeV. Electron driven PWFA has validated the viability of an energy doubling of 42 GeV, which sheds light on the promising application of plasma-based accelerators towards high energy physics. Nevertheless, the obtainable energy gain is intrinsically liable to the drive energy contents. To accelerate particles to the energy frontier, multi-staging scheme is necessary which boosts the particle energy stage by stage. Given the limited energies contained in the existing lasers or electron bunches, tens or hundreds of stages are required, which makes this scheme essentially challenging.

Proton driven PWFA stands out under this circumstance as existing proton bunches possess huge energies and large particle populations. The energy contents are hundreds to thousands times more than those in the laser or electron drivers. Therefore, proton bunches are the most promising candidates to accelerate particles to the TeV-level in a single plasma stage. This scheme was proposed nearly a decade ago assuming a short proton bunch, which is not available in practice at present. But it has received considerable attention with the emergence of the SSM of long proton bunches, which enables strong wakefield excitation. Also the success of the AWAKE experiment, which is dedicated to the electron acceleration in the long proton bunch driven plasma wakefield, further promotes the faith of scientists on this scheme. In this thesis, by means of simulations, we have investigated the wakefield and acceleration characteristics driven by a single or multiple short proton bunches which might be produced in the future and also by the long proton bunches which are more realistic.

Unlike the laser or electron drivers, the proton bunches with positive charges will attract the background plasma electrons instead of repelling them, which causes the adverse "phase mixing" effect (introduced in Chap. 3). There are nonuniform plasma electrons distributing in the accelerating region, that is, a "clean" plasma bubble (a pure ion column) doesn't stand any more. A rough view of this can be found in Fig. 1.3 where the driver is positively charged. The resultant radially- and time- varying transverse focusing dilutes the accelerated beam, leading to considerable growth of the beam radius and beam emittance. The large beam emittance will further blow up the beam during the beam transport. It follows that the deterioration of beam quality makes it unsuitable for the applications in the colliders as the collision luminosity $\mathcal{L}$ is inversely proportional to the product of transverse beam sizes [88]:

$$\mathcal{L} = f\frac{N_1 N_2}{4\pi\sigma_\mathrm{x}\sigma_\mathrm{y}}. \tag{1.5}$$

Here f is the collision frequency, N_1 and N_2 are the particle populations of two gaussian beams colliding head-on, σ_x and σ_y are the horizontal and vertical beam sizes at the interaction point assuming equal beams. In this thesis, by means of nonuniform plasma, we are devoted to producing ultra-high energy and high quality electrons and positrons, which are the promising candidates for the future energy frontier lepton colliders.

1.3.2 Outline

In this thesis, we present the motivation, physics and particle-in-cell simulation results of proton driven plasma wakefield acceleration. The aim is to demonstrate the viability of generating high energy high quality electrons and positrons, which are pursued by future energy frontier lepton colliders.

In Chap. 2, we provide a theoretical description of the beam-driven plasma wakefield acceleration in both the linear and nonlinear regime. We demonstrate the beam loading effect and the associated properties of the beam. We introduce the algorithms of both fully relativistic and the quasi-static particle-in-cell codes, the latter of which is employed in our major simulations.

In Chap. 3, we introduce a hollow plasma channel into a single short proton bunch driven scheme for electron acceleration. A favorable accelerating structure is formed with a strong and radially uniform accelerating field, and is completely free from plasma electrons and ions (i.e., no transverse plasma wakefields). In addition, it preserves its shape and location up to driver depletion, thus providing a high acceleration efficiency. As a result, the witness electron bunch carrying the charge of about 10% of 1 TeV proton driver charge is accelerated to 0.6 TeV over a single hollow plasma channel of 700 m. More importantly, its normalized emittance is preserved.

In Chap. 4, we present the simulation results of electron acceleration driven by a train of short, lower charge proton bunches instead of a single one, which is more challenging to obtain by a significant length compression. With a hollow plasma channel to remove the defocusing from the plasma ions, the multiple proton bunches are well confined by the basin-like potential well, which is formed by the plasma electrons attracted within the channel. As a result, the proton bunches can stably operate in the blowout regime, resonantly exciting strong plasma waves. The witness electrons are accelerated to 0.47 TeV over a plasma of length 150 m. The hollow channel also brings beneficial features for the accelerated beam, such as emittance preservation and low energy spread (1.3%).

In Chap. 5, we present the simulation results of positron acceleration driven by similarly multiple proton bunches in a hollow plasma channel. The acceleration of positrons is different because the accelerating structure is strongly charge dependent. We propose to load an extra electron bunch to repel the undesired plasma electrons and meanwhile reduce the plasma density slightly to shift the accelerating phase with a conducive slope to the positron bunch. In this way, the positron bunch can be accelerated to 0.4 TeV (40% of the driver energy) with an energy spread as low as 1% and well preserved normalized emittance.

In Chap. 6, we assess the adverse transverse effects induced by the misalignment between the beam and the hollow plasma channel by simulating an initially offset or tilted driving proton bunch. It shows that the proton bunch is less prone to the initial misalignment as it is strongly focused by the hollow channel, while the induced asymmetric transverse field leads to the drastic deflection and eventually loss of the witness electron bunch. Adopting a near-hollow plasma is promising to mitigate the issue, though further work is needed to determine whether the hollow plasma structure is suitable for future applications.

In Chap. 7, we simulate the self-modulated long proton bunch in uniform plasma. We propose a slightly sophisticated plasma taper to compensate the adverse effect of huge proton loss caused by the significant decrease of wake phase velocity during the development of the seeded self-modulation. We finally maintain 24% of the initial beam charge after micro-bunching and boost the wakefield amplitude by 30%.

In Chap. 8, we summarise the main simulation results and discuss future prospects of proton-driven plasma wakefield acceleration.

References

1. Aad G, Abajyan T, Abbott B, Abdallah J, Khalek SA, Abdelalim A, Abdinov O, Aben R, Abi B, Abolins M et al (2012) Observation of a new particle in the search for the standard model higgs boson with the ATLAS detector at the LHC. Phys Lett B 716(1):1–29
2. Chatrchyan S, Khachatryan V, Sirunyan AM, Tumasyan A, Adam W, Aguilo E, Bergauer T, Dragicevic M, Ero J, Fabjan C et al (2012) Observation of a new boson at a mass of 125 GeV with the CMS experiment at the LHC. Phys Lett B 716(1):30–61
3. Aicheler M, Burrows P, Draper M, Garvey T, Lebrun P, Peach K, Phinney N, Schmickler H, Schulte D, Toge N (2014) A multi-TeV linear collider based on CLIC technology: CLIC conceptual design report. Technical Report, SLAC national accelerator lab., Menlo Park, CA (United States)
4. Humphries S (2013) Principles of charged particle acceleration. Courier Corporation
5. Lee S-Y (2011) Accelerator physics. World Scientific Publishing Company, Singapore
6. Adolphsen C, BaroneM, Barish B, Buesser K, Burrows P, Carwardine J, Clark J, Durand HM, Dugan G, Elsen E et al (2013) The international linear collider technical design report-volume 3. II: Accelerator baseline design. arXiv:1306.6328
7. Fitzpatrick R (1998) Introduction to plasma physics, Lecture notes
8. Tajima T, Dawson J (1979) Laser electron accelerator. Phys Rev Lett 43(4):267
9. Esarey E, Sprangle P, Krall J, Ting A (1996) Overview of plasma-based accelerator concepts. IEEE Trans Plasma Sci 24(2):252–288
10. Andreev N (1992) Resonant excitation of wakefields by a laser pulse. JETP lett 55(10):571–576
11. Krall J, Ting A, Esarey E, Sprangle P (1993) Enhanced acceleration in a self-modulated-laser wake-field accelerator. Phys Rev E 48(3):2157
12. Strickland D, Mourou G (1985) Compression of amplified chirped optical pulses. Opt Commun 55(6):447–449
13. Maine P, Strickland D, Bado P, Pessot M, Mourou G (1988) Generation of ultrahigh peak power pulses by chirped pulse amplification. IEEE J Quantum Electron 24(2):398–403
14. Mourou GA, Barty C, Perry MD (1998) Ultrahigh-intensity lasers: physics of the extreme on a tabletop. Phys Today 51(1):22–28
15. Mora P, Antonsen TM Jr (1996) Electron cavitation and acceleration in the wake of an ultraintense, self-focused laser pulse. Phys Rev E 53(3):R2068
16. Pukhov A, Meyer-ter Vehn J (2002) Laser wake field acceleration: the highly non-linear brokenwave regime. Appl Phys B 74(4–5):355–361
17. Faure J, Glinec Y, Pukhov A, Kiselev S, Gordienko S, Lefebvre E, Rousseau J-P, Burgy F, Malka V (2004) A laser plasma accelerator producing monoenergetic electron beams. Nature 431(7008):541
18. Geddes C, Toth C, Van Tilborg J, Esarey E, Schroeder C, Bruhwiler D, Nieter C, Cary J, Leemans W (2004) High-quality electron beams from a laser wakefield accelerator using plasmachannel guiding. Nature 431(7008):538

19. Mangles SP, Murphy C, Najmudin Z, Thomas AGR, Collier J, Dangor AE, Divall E, Foster P, Gallacher J, Hooker C et al (2004) Monoenergetic beams of relativistic electrons from intense laser plasma interactions. Nature 431(7008):535
20. Hooker SM (2013) Developments in laser-driven plasma accelerators. Nat Photonics 7(10):775
21. Leemans WP, Nagler B, Gonsalves AJ, Toth C, Nakamura K, Geddes CG, Esarey E, Schroeder C, Hooker S (2006) GeV electron beams from a centimetre-scale accelerator. Nat Phys 2(10):696
22. Leemans W, Gonsalves A, Mao H-S, Nakamura K, Benedetti C, Schroeder C, Toth C, Daniels J, Mittelberger D, Bulanov S et al (2014) Multi-GeV electron beams from capillary-discharge-guided subpetawatt laser pulses in the self-trapping regime. Phys Rev Lett 113(24):245002
23. Esarey E, Schroeder C, Leemans W (2009) Physics of laser-driven plasmabased electron accelerators. Rev Mod Phys 81(3):1229
24. Rosenbluth M, Liu C (1972) Excitation of plasma waves by two laser beams. Phys Rev Lett 29(11):701
25. Chiou T, Katsouleas T, Decker C, Mori W, Wurtele J, Shvets G, Su J (1995) Laser wake-field acceleration and optical guiding in a hollow plasma channel. Phys Plasmas 2(1):310–318
26. Cros B (2016) Plasma physics: compact coupling for a two-stage accelerator. Nature 530(7589):165
27. Steinke S, Van Tilborg J, Benedetti C, Geddes C, Schroeder C, Daniels J, Swanson K, Gonsalves A, Nakamura K, Matlis N et al (2016) Multistage coupling of independent laser-plasma accelerators. Nature 530(7589):190
28. Kudryavtsev A, Lotov K, Skrinsky A (1998) Plasma wake-field acceleration of high energies: physics and perspectives. Nucl Instrum Methods Phys Res Sect A 410(3):388–395
29. Rosenzweig J, Barov N, Murokh A, Colby E, Colestock P (1998) Towards a plasma wake-field acceleration-based linear collider. Nucl Instrum Methods Phys Res Sect A 410(3):532–543
30. Shiltsev VD (2012) High-energy particle colliders: past 20 years, next 20 years, and beyond. Physics-Uspekhi 55(10):965
31. Chen P, Dawson J, Hufi RW, Katsouleas T (1985) Acceleration of electrons by the interaction of a bunched electron beam with a plasma. Phys Rev Lett 54(7):693
32. Lee S, Katsouleas T, Hemker R, Dodd E, Mori W (2001) Plasmawake field acceleration of a positron beam. Phys Rev E 64(4):045501
33. Rosenzweig J (1987) Nonlinear plasma dynanics in the plasma wakefield accelerator. IEEE Trans Plasma Sci 15(2):186–191
34. Rosenzweig J, Breizman B, Katsouleas T, Su J (1991) Acceleration and focusing of electrons in two-dimensional nonlinear plasma wake fields. Phys Rev A 44(10):R6189
35. Ruth RD, Morton P, Wilson PB, Chao A (1984) A plasma wake field accelerator. Part Accel 17(SLAC-PUB-3374):171
36. Bane KL, Chen P, Wilson P (1985) Collinear wake field acceleration. Technical Report, Stanford Linear Accelerator Center, CA (USA)
37. Katsouleas TC, Wilks S, Chen P, Dawson JM, Su JJ (1987) Beam loading in plasma accelerators. Part Accel 22:81–99
38. Hogan M, Clayton C, Huang C, Muggli P, Wang S, Blue B, Walz D, Marsh K, O'Connell C, Lee S et al (2003) Ultrarelativistic-positron-beam transport through meter-scale plasmas. Phys Rev Lett 90(20):205002
39. Muggli P, Blue B, Clayton C, Deng S, Decker F-J, Hogan M, Huang C, Iverson R, Joshi C, Katsouleas T et al (2004) Meter-scale plasma-wake field accelerator driven by a matched electron beam. Phys Rev Lett 93(1):014802
40. Lu W, Huang C, Zhou M, Mori W, Katsouleas T (2005) Limits of linear plasma wakefield theory for electron or positron beams. Phys Plasmas 12(6):063101
41. Muggli P, Hogan MJ (2009) Review of high-energy plasma wake field experiments. C R Phys 10(2–3):116–129
42. Blumenfeld I, Clayton CE, Decker F-J, Hogan MJ, Huang C, Ischebeck R, Iverson R, Joshi C, Katsouleas T, Kirby N et al (2007) Energy doubling of 42 GeV electrons in a metre-scale plasma wakefield accelerator. Nature 445(7129):741

43. Litos M, Adli E, An W, Clarke C, Clayton C, Corde S, Delahaye J, England R, Fisher A, Frederico J et al (2014) High-efficiency acceleration of an electron beam in a plasma wake field accelerator. Nature 515(7525):92
44. Corde S, Adli E, Allen J, An W, Clarke C, Clayton C, Delahaye J, Frederico J, Gessner S, Green S et al (2015) Multi-gigaelectronvolt acceleration of positrons in a self-loaded plasma wakefield. Nature 524(7566):442
45. Blue BE, Clayton C, O'connell C, Decker F-J Hogan M, Huang C, Iverson R, Joshi C, Katsouleas T, Lu W et al (2003) Plasma-wakefield acceleration of an intense positron beam. Phys Rev Lett 90(21):214801
46. Hogan M, Barnes C, Clayton C, Decker F, Deng S, Emma P, Huang C, Iverson R, Johnson D, Joshi C et al (2005) Multi-GeV energy gain in a plasma-wake field accelerator. Phys Rev Lett 95(5):054802
47. Wang X, Muggli P, Katsouleas T, Joshi C, Mori W, Ischebeck R, Hogan M (2009) Optimization of positron trapping and acceleration in an electronbeam-driven plasma wakefield accelerator. Phys Rev ST Accel Beams 12(5):051303
48. Lee S, Katsouleas T, Muggli P, Mori W, Joshi C, Hemker R, Dodd E, Clayton C, Marsh K, Blue B et al (2002) Energy doubler for a linear collider. Phys Rev ST Accel Beams 5(1):011001
49. Chen P, Su J, Dawson J, Bane KLF, Wilson P (1986) Energy transfer in the plasma wake-field accelerator. Phys Rev Lett 56(12):1252
50. Farmer J, Martorelli R, Pukhov A (2015) Transformer ratio saturation in a beam-driven wakefield accelerator. Phys Plasmas 22(12):123113
51. Caldwell A, Adli E, Amorim L, Apsimon R, Argyropoulos T, Assmann R, Bachmann A-M, Batsch F, Bauche J, Olsen VB et al (2016) Path to AWAKE: evolution of the concept. Nucl Instrum Methods Phys Res Sect A 829:3–16
52. Danson C, Hillier D, Hopps N, Neely D (2015) Petawatt class lasers worldwide. High Power Laser Sci Eng 3:e3
53. Nakajima K, Deng A, Zhang X, Shen B, Liu J, Li R, Xu Z, Ostermayr T, Petrovics S, Klier C et al (2011) Operating plasma density issues on large-scale laser-plasma accelerators toward high-energy frontier. Phys Rev ST Accel Beams 14(9):091301
54. Schroeder C, Esarey E, Leemans W (2012) Beamstrahlung considerations in laser-plasma-accelerator-based linear colliders. Phys Rev ST Accel Beams 15(5):051301
55. Schroeder C, Benedetti C, Esarey E, Leemans W (2016) Laser-plasmabased linear collider using hollow plasma channels. Nucl Instrum Methods Phys Res Sect A 829:113–116
56. Caldwell A, Lotov K, Pukhov A, Simon F (2009) Proton-driven plasmawake field acceleration. Nat Phys 5(5):363
57. Lotov K (2010) Simulation of proton driven plasma wakefield acceleration. Phys Rev ST Accel Beams 13(4):041301
58. Yi L, Shen B, Lotov K, Ji L, Zhang X, Wang W, Zhao X, Yu Y, Xu J, Wang X et al (2013) Scheme for proton-driven plasma-wake field acceleration of positively charged particles in a hollow plasma channel. Phys Rev ST Accel Beams 16(7):071301
59. Yi L, Shen B, Ji L, Lotov K, Sosedkin A, Wang W, Xu J, Shi Y, Zhang L, Xu Z et al (2014) Positron acceleration in a hollow plasma channel up to TeV regime Sci Rep 4:4171
60. Leemans W, Duarte R, Esarey E, Fournier S, Geddes C, Lockhart D, Schroeder C, Tfioth C, Vay J-L, Zimmermann S (2010) The BErkeley lab laser accelerator (BELLA): a 10 GeV laser plasma accelerator. AIP Conf Proc AIP 1299:3–11
61. Xia G, Caldwell A, Lotov K, Pukhov A, Kumar N, An W, Lu W, Mori W, Joshi C, Huang C et al (2010) Update of proton driven plasma wake-field acceleration. AIP Conf Proc AIP 1299:510–515
62. Caldwell A, Lotov K, Pukhov A, Xia G (2010) Plasma wake field excitation with a 24 GeV proton beam. Plasma Phys Control Fusion 53(1):014003
63. Caldwell A, Lotov K (2011) Plasma wakefield acceleration with a modulated proton bunch. Phys Plasmas 18(10):103101
64. Lotov K (1998) Instability of long driving beams in plasma wake field accelerators. In: Proceedings of the 6th European particle accelerator conference, Stockholm, pp 806–808

65. Kumar N, Pukhov A, Lotov K (2010) Self-modulation instability of a long proton bunch in plasmas. Phys Rev Lett 104(25):255003
66. Krall J, Joyce G (1995) Transverse equilibrium and stability of the primary beam in the plasma wake-field accelerator. Phys Plasmas 2(4):1326–1331
67. Whittum DH (1997) Transverse two-stream instability of a beam with a bennett profile. Phys Plasmas 4(4):1154–1159
68. Lotov K (2011) Controlled self-modulation of high energy beams in a plasma. Phys Plasmas 18(2):024501
69. Lotov K (2015) Effect of beam emittance on self-modulation of long beams in plasma wakefield accelerators. Phys Plasmas 22(12):123107
70. Lotov K (2015) Physics of beam self-modulation in plasma wake field accelerators. Phys Plasmas 22(10):103110
71. Schroeder C, Benedetti C, Esarey E, Gruner F, Leemans W (2012) Coupled beam hose and self-modulation instabilities in overdense plasma. Phys Rev E 86(2):026402
72. Schroeder C, Benedetti C, Esarey E, Gruner F, Leemans W (2013) Coherent seeding of self-modulated plasma wakefield accelerators. Phys Plasmas 20(5):056704
73. Vieira J, Mori W, Muggli P (2014) Hosing instability suppression in selfmodulated plasma wakefields. Phys Rev Lett 112(20):205001
74. Assmann R, Bingham R, Bohl T, Bracco C, Buttenschon B, Butterworth A, Caldwell A, Chattopadhyay S, Cipiccia S, Feldbaumer E et al (2014) Proton-driven plasma wakefield acceleration: a path to the future of high-energy particle physics. Plasma Phys Control Fusion 56(8):084013
75. Bracco C, Gschwendtner E, Petrenko A, Timko H, Argyropoulos T, Bartosik H, Bohl T, Muller JE, Goddard B, Meddahi M et al (2014) Beam studies and experimental facility for the AWAKE experiment at CERN. Nucl Instrum Methods Phys Res Sect A 740:48–53
76. Gschwendtner E, Adli E, Amorim L, Apsimon R, Assmann R, Bachmann A-M, Batsch F, Bauche J, Olsen VB, Bernardini M et al (2016) AWAKE, the advanced proton driven plasma wakefield acceleration experiment at CERN. Nucl Instrum Methods Phys Res Sect A 829:76–82
77. Turner M et al (AWAKE Collaboration) (2019) Experimental observation of plasma wakefield growth driven by the seeded self-modulation of a proton bunch. Phys Rev Lett 122(5):054801
78. Adli E et al (AWAKE Collaboration) (2019) Experimental observation of proton bunch modulation in a plasma, at varying plasma densities. Phys Rev Lett 122(5):054802
79. Adli E, Ahuja A, Apsimon O, Apsimon R, Bachmann A-M, Barrientos D, Batsch F, Bauche J, Olsen VB, Bernardini M et al (2018) Acceleration of electrons in the plasma wakefield of a proton bunch. Nature 561(7723):363
80. Fang Y, Vieira J, Amorim L, Mori W, Muggli P (2014) The effect of plasma radius and profile on the development of self-modulation instability of electron bunches. Phys Plasmas 21(5):056703
81. Fang Y, Yakimenko V, Babzien M, Fedurin M, Kusche K, Malone R, Vieira J, Mori W, Muggli P (2014) Seeding of self-modulation instability of a long electron bunch in a plasma. Phys Rev Lett 112(4):045001
82. Vieira J, Fang Y, Mori W, Silva L, Muggli P (2012) Transverse selfmodulation of ultra-relativistic lepton beams in the plasma wake field accelerator. Phys Plasmas 19(6):063105
83. Gross M, Brinkmann R, Good J, Gruner F, Khojoyan M, de la Ossa AM, Osterhofi J, Pathak G, Schroeder C, Stephan F (2014) Preparations for a plasma wakefield acceleration (PWA) experiment at PITZ. Nucl Instrum Methods Phys Res Sect A 740:74–80
84. Lishilin O, Gross M, Brinkmann R, Engel J, Gruner F, Koss G, Krasilnikov M, de la Ossa AM, Mehrling T, Osterhofi J et al (2016) First results of the plasma wake field acceleration experiment at PITZ. Nucl Instrum Methods Phys Res Sect A 829:37–42
85. Adli E, Olsen VB, Lindstrfim C, Muggli P, Reimann O, Vieira J, Amorim L, Clarke C, Gessner S, Green S et al (2016) Progress of plasma wake-field self-modulation experiments at FACET. Nucl Instrum Methods Phys Res Sect A 829:334–338
86. Apsimon OM, Burt G, Hanahoe K, Xia G Hidding B (2010) iMPACT, undulator-based multi-bunch plasma accelerator. In: Proceedings of international particle accelerator conference (IPAC), Busan, Korea, p 2609

87. Pompili R, Anania M, Bellaveglia M, Biagioni A, Bisesto F, Chiadroni E, Cianchi A, Croia M, Curcio A, Di Giovenale D et al (2016) Beam manipulation with velocity bunching for PWFA applications. Nucl Instrum Methods Phys Res Sect A 829:17–23
88. Herr W, Muratori B (2006) Concept of luminosity. CERN Accelerator School Intermediate accelerator physics, p 361

Chapter 2
Physics of Plasma Wakefield Acceleration in Uniform Plasma

2.1 Introduction

Plasma wakefield acceleration as an advanced accelerator concept has attracted enormous attention in the last decades. By virtue of strong plasma wave fields excited by the driving bunch, a trailing bunch can be accelerated to an ultrahigh energy in a short distance. In this chapter, we provide a theoretical description of the plasma wakefield acceleration and elucidate the underlying physics. First of all, following the insights from Ruth [1] and Katsouleas [2], we demonstrate the plasma response and the characteristics of the transverse and longitudinal wakefields in the linear regime where the plasma is overdense. After discussing the limit of the linear theory [3], we then describe the nonlinear "blowout" regime for underdense plasma using a brief sheath model developed by Lu [4, 5]. Next, we discuss the beam loading in both linear [2, 6] and nonlinear [7, 8] regimes for the sake of flat wakefields and small energy spread, and introduce associated physical properties [1] like normalized emittance, transformer ratio, energy transfer efficiency, etc. In the end, we demonstrate the simulation algorithm of the particle-in-cell (PIC) codes, and compare the full relativistic PIC code and the quasi-static code which serves most of our self-consistent simulations in this thesis due to the high computing efficiency.

2.2 Linear Theory in Nonrelativistic Plasma

2.2.1 *Plasma Density Response*

In the linear regime, assuming an axisymmetric, ultrarelativistic ($v_b \sim c$) electron beam traverses into a cold, unmagnetized plasma and the beam density (n_b) is much less than the initial plasma density (n_0), the plasma response to the beam can be described with fluid equations following the treatment in Refs. [1, 2]. The plasma

Y. Li, *Studies of Proton Driven Plasma Wakefield Acceleration*, Springer Theses,
https://doi.org/10.1007/978-3-030-50116-7_2

density change in a local region is associated with plasma flowing into or out of this region via the continuity equation

$$\frac{\partial n}{\partial t} = -\nabla \cdot (n\boldsymbol{v}). \tag{2.1}$$

The quantity $n(r, z, t)$ refers to the dynamic plasma electron density in the cylindrical coordinate system and $\boldsymbol{v}$ is the plasma electron velocity. The massive plasma ions are considered stationary in the discussed time scale with the uniform number density n_0. The motion of the plasma fluid in response to the fields complies with the Lorentz force law

$$\frac{\partial \boldsymbol{v}}{\partial t} + (\boldsymbol{v} \cdot \nabla)\boldsymbol{v} = -\frac{e}{m_\mathrm{e}}\left(\boldsymbol{E} + \frac{\boldsymbol{v} \times \boldsymbol{B}}{c}\right), \tag{2.2}$$

Here m_e is the electron mass, e is the elementary charge ($-e$ for an electron and $+e$ for a positron), and c is the speed of light. The fields evolve according to Maxwell's equations, which are given in Gaussian units:

$$\nabla \times \boldsymbol{B} = \frac{4\pi}{c}\boldsymbol{J} + \frac{1}{c}\frac{\partial \boldsymbol{E}}{\partial t}, \tag{2.3}$$

$$\nabla \times \boldsymbol{E} = -\frac{1}{c}\frac{\partial \boldsymbol{B}}{\partial t}, \tag{2.4}$$

$$\nabla \cdot \boldsymbol{E} = 4\pi\rho, \tag{2.5}$$

$$\nabla \cdot \boldsymbol{B} = 0. \tag{2.6}$$

Since the beam is ultrarelativistic, we can consider it unaltered or "rigid" during the evolution of the plasma fields. That is, the beam current remains fixed and the radial motion of the beam dynamics in response to plasma fields is ignored. As $n_\mathrm{b} \ll n_0$, the plasma density perturbation n_1 is significantly small compared to n_0, where $n = n_0 + n_1$. It follows that the plasma velocity $\boldsymbol{v}$, and fields $\boldsymbol{E}, \boldsymbol{B}$ are all first order perturbations. Keeping only linear terms, Eqs. 2.1 and 2.2 can be written as

$$-\frac{1}{n_0}\frac{\partial n_1}{\partial t} = \nabla \cdot \boldsymbol{v}, \tag{2.7}$$

$$\frac{\partial \boldsymbol{v}}{\partial t} = -\frac{e\boldsymbol{E}}{m_\mathrm{e}}. \tag{2.8}$$

Apparently, the term $\frac{\boldsymbol{v}\times\boldsymbol{B}}{c}$ in Eq. 2.2 appears in second order. Taking the divergence of Eq. 2.8 and plugging the expression in 2.7 for $\nabla \cdot \boldsymbol{v}$ into it gives

$$\frac{\partial^2 n_1}{\partial t^2} = \frac{en_0}{m_\mathrm{e}}\nabla \cdot \boldsymbol{E}. \tag{2.9}$$

According to Gauss's law,

$$\nabla \cdot \boldsymbol{E} = -4\pi e n_1 - 4\pi e n_{\mathrm{b}}, \tag{2.10}$$

then by combining Eqs. 2.9 and 2.10 we can obtain the second-order differential equation of the plasma density perturbation

$$\frac{\partial^2 n_1}{\partial t^2} + \omega_{\mathrm{p}}^2 n_1 = -\omega_{\mathrm{p}}^2 n_{\mathrm{b}}, \tag{2.11}$$

where $\omega_{\mathrm{p}} = \sqrt{4\pi e^2 n_0 / m_{\mathrm{e}}}$ is the plasma electron frequency.

Given an arbitrary ultrarelativistic, axisymmetric bunch with the density form consisting of the longitudinal and transverse components

$$n_{\mathrm{b}}(r, z, t) = R(r) Z(z - ct), \tag{2.12}$$

and applying the combined coordinate $\xi = z - ct$, the differential Eq. 2.11 can be expressed as

$$\frac{\partial^2 n_1(r, \xi)}{\partial \xi^2} + k_{\mathrm{p}}^2 n_1(r, \xi) = -k_{\mathrm{p}}^2 R(r) Z(\xi), \tag{2.13}$$

where $k_{\mathrm{p}} = \omega_{\mathrm{p}}/c$ is the plasma wavenumber, and ξ represents the relative position to the bunch head. From this equation, we can learn that the radial distribution of the plasma density perturbation exactly follows the radial profile of the driving bunch density, which can be validated by Fig. 2.1a shown later.

The solution to Eq. 2.13 without source is

$$n_1(r, \xi) = a \sin k_{\mathrm{p}}\xi + b \cos k_{\mathrm{p}}\xi. \tag{2.14}$$

Assuming the source is a point source following a Dirac delta function, i.e., $Z(\xi) = \delta(\xi)$, and integrating both sides of Eq. 2.13 over $\xi = 0$, we find

$$\frac{\partial n_1(r, \xi)}{\partial \xi} \Big|_{\xi=0} = -k_{\mathrm{p}}^2 R(r), \tag{2.15}$$

thus $a = -k_{\mathrm{p}} R(r)$. In addition, b is equal to 0 as the slope of n_1 at $\xi = 0$ is negative. We can then write the Green's function solution to Eq. 2.13 as

$$G_{n_1}(r, \xi) = -k_{\mathrm{p}} R(r) \sin(k_{\mathrm{p}}\xi) H(-\xi), \tag{2.16}$$

where $H(-\xi)$ is the Heaviside step function with the value of 1 when $\xi < 0$ and the value of 0 when $\xi > 0$. This is consistent with the fact that the region $\xi < 0$ is behind the bunch source where there is plasma perturbation, while for $\xi > 0$, there is no source passing by and thus no plasma perturbation.

The plasma response to an arbitrary bunch distribution $n_{\mathrm{b}}(r, \xi) = R(r)Z(\xi)$ can be built up point by point by integrating the Green's function against the bunch source, which gives

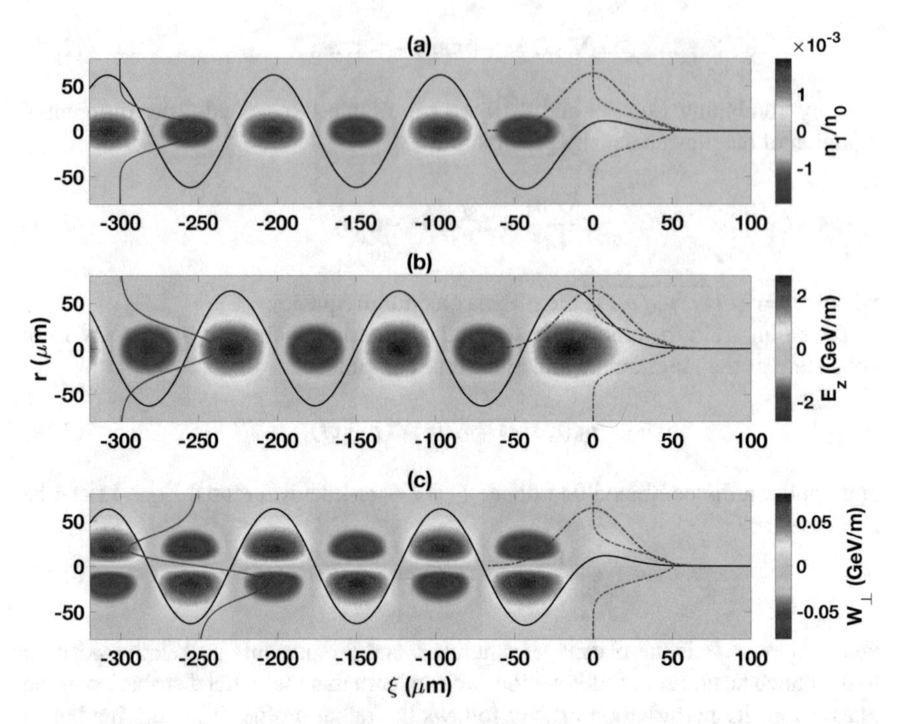

Fig. 2.1 Spatial distribution of (**a**) the perturbed plasma electron density, (**b**) excited longitudinal electric fields and (**c**) transverse plasma fields when an electron beam propagates into a uniform overdense plasma. The blue and magenta dashed lines indicate the longitudinal and radial profiles of the driving beam, respectively. The red lines show transverse lineouts of the corresponding plasma density and fields. The black lines in (**a**) and (**b**) are the longitudinal on-axis lineouts while the black one in (**c**) gives the off-axis lineout. Because the radial force on-axis is zero as indicated by the red line in (**c**). The black lines facilitate the comparison of the oscillation periods and phases of the plasma density and wakefields

$$n_1(r, \xi) = -k_p R(r) \int_{-\infty}^{\xi} \sin k_p(\xi - \xi') Z(\xi') d\xi'. \tag{2.17}$$

Figure 2.1a shows the plasma density perturbation in response to a bi-Gaussian drive beam with $n_b(r, \xi) = \frac{N_b}{(2\pi)^{3/2}\sigma_z\sigma_r^2} e^{-\frac{r^2}{2\sigma_r^2} - \frac{\xi^2}{2\sigma_z^2}}$ and $n_b \ll n_0$. N_b is the beam population, σ_r and σ_z are the RMS beam radius and length, respectively. The beam is assumed an electron beam with the following parameters: $N_b = 1 \times 10^9$, $\sigma_r = 15\,\mu\text{m}$, $\sigma_z = 20\,\mu\text{m}$. The plasma density is $1 \times 10^{17}\,\text{cm}^{-3} \sim 10 n_b(0, 0)$. We see the perturbed plasma oscillates sinusoidally at the plasma frequency in time (see the black line) and follows the radial density distribution of the electron beam in space (compare the red line and the magenta dashed line).

2.2.2 Plasma Fields

Previous derivation has demonstrated the plasma density change in time and space in response to a drive bunch. In this part, we will derive the generated plasma fields due to beam disturbance so that we can in turn evaluate their effect on the beam.

Taking the curl of both sides of Eq. 2.4, then applying the relation $\nabla \times (\nabla \times \boldsymbol{E}) = \nabla(\nabla \cdot \boldsymbol{E}) - \nabla^2 \boldsymbol{E}$ to its left side and plugging the expression for $\nabla \times \boldsymbol{B}$ in Eq. 2.3 into its right side, after rearrangement we can get

$$\nabla^2 \boldsymbol{E} - \frac{1}{c^2}\frac{\partial^2 \boldsymbol{E}}{\partial t^2} = \frac{4\pi}{c^2}\frac{\partial \boldsymbol{J}}{\partial t} + 4\pi \nabla \rho. \tag{2.18}$$

The Laplacian operator can be separated into the transverse and longitudinal components, i.e., $\nabla^2 = \nabla_\perp^2 + \frac{\partial^2}{\partial z^2}$. Assuming the plasma fields only depend on the combined coordinate $\xi = z - ct$, then $\frac{\partial^2}{\partial z^2} = \frac{1}{c^2}\frac{\partial^2}{\partial t^2}$. The source comes from the drive beam and the plasma, where $\boldsymbol{J} = \boldsymbol{J}_\mathbf{b} + \boldsymbol{J}_\mathbf{p}$ with $\boldsymbol{J}_\mathbf{b} = c\rho_\mathrm{b}\hat{z}$ and $\rho = \rho_\mathrm{b} + \rho_\mathrm{p}$. From Eq. 2.8, we get a relation between $\boldsymbol{J}_\mathbf{p}$ and $\boldsymbol{E}$ by multiplying both sides by $-en_0$

$$\frac{\partial \boldsymbol{J}_\mathbf{p}}{\partial t} = \frac{\omega_\mathrm{p}^2}{4\pi}\boldsymbol{E}. \tag{2.19}$$

Plugging this into Eq. 2.18 gives

$$(\nabla_\perp^2 - k_\mathrm{p}^2)\boldsymbol{E} = \frac{4\pi}{c}\frac{\partial \rho_\mathrm{b}}{\partial t}\hat{z} + 4\pi\nabla(\rho_\mathrm{b} + \rho_\mathrm{p}). \tag{2.20}$$

By directly applying Faraday's Law (see Eq. 2.4) into this equation, the magnetic field can be expressed as

$$(\nabla_\perp^2 - k_\mathrm{p}^2)\boldsymbol{B} = -4\pi\nabla \times \rho_\mathrm{b}\hat{z}. \tag{2.21}$$

Then we have the relations among field components

$$(\nabla_\perp^2 - k_\mathrm{p}^2)E_z = 4\pi\frac{\partial \rho_\mathrm{p}}{\partial z}, \tag{2.22}$$

$$(\nabla_\perp^2 - k_\mathrm{p}^2)E_\mathrm{r} = 4\pi\frac{\partial}{\partial r}(\rho_\mathrm{b} + \rho_\mathrm{p}), \tag{2.23}$$

$$(\nabla_\perp^2 - k_\mathrm{p}^2)B_\theta = 4\pi\frac{\partial \rho_\mathrm{b}}{\partial r}. \tag{2.24}$$

Here we have used the relation $\frac{\partial}{\partial z} = -\frac{1}{c}\frac{\partial}{\partial t}$. Denoting the longitudinal and transverse plasma field functions as $\boldsymbol{W}_\parallel$ and $\boldsymbol{W}_\perp$ where

$$\boldsymbol{W}_{\parallel,\perp} = \left(\boldsymbol{E} + \frac{\boldsymbol{v}_\mathbf{b} \times \boldsymbol{B}}{c}\right)_{z,r}, \tag{2.25}$$

it gives after employing the fact that the beam is ultrarelativistic ($\boldsymbol{v}_\mathrm{b} = c\hat{z}$),

$$(\nabla_\perp^2 - k_\mathrm{p}^2) W_\parallel = 4\pi \frac{\partial \rho_\mathrm{p}}{\partial z}, \tag{2.26}$$

$$(\nabla_\perp^2 - k_\mathrm{p}^2) W_\perp = 4\pi \frac{\partial \rho_\mathrm{p}}{\partial r}, \tag{2.27}$$

where $W_\parallel = E_\mathrm{z}$ and $W_\perp = E_\mathrm{r} - B_\theta$. Equation 2.27 shows that the beam source contributing to E_r and B_θ cancels out at the radial force $W_\perp$, which is therefore entirely sourced by the plasma perturbation. This coincides with the fact that, in vacuum the repulsive space charge force of an ultrarelativistic beam is exactly canceled by its focusing magnetic force generated by the self-current. Equation 2.26 indicates the same rule to the longitudinal field.

Assuming the beam source is a point charge $n_\mathrm{b}(r, \xi) = \delta(r)\delta(\xi)$, the plasma density perturbation is $n_1(r, \xi) = -k_\mathrm{p}\delta(r)\sin(k_\mathrm{p}\xi)H(-\xi)$ according to Eq. 2.16. Then Eq. 2.22 becomes

$$(\nabla_\perp^2 - k_\mathrm{p}^2) E_\mathrm{z} = 4\pi e k_\mathrm{p}^2 \delta(r)\cos(k_\mathrm{p}\xi)H(-\xi). \tag{2.28}$$

In a cylindrical system, the operator $\nabla_\perp^2$ is expressed as

$$\nabla_\perp^2 = \frac{1}{r}\frac{\partial}{\partial r}\left(r\frac{\partial}{\partial r}\right) + \frac{1}{r^2}\frac{\partial^2}{\partial \theta^2}. \tag{2.29}$$

The source is azimuthally symmetric, thus the derivatives with respect to θ are zero and Eq. 2.28 is expanded to

$$\frac{1}{r}\frac{\partial}{\partial r}\left(r\frac{\partial E_\mathrm{z}}{\partial r}\right) - k_\mathrm{p}^2 E_\mathrm{z} = 4\pi e k_\mathrm{p}^2 \delta(r)\cos(k_\mathrm{p}\xi)H(-\xi). \tag{2.30}$$

The Green's function solution to this inhomogeneous Helmholtz equation with a radial delta source is [9]

$$G_{E_\mathrm{z}}(r, \xi) = -2ek_\mathrm{p}^2 K_0(k_\mathrm{p}r)\cos(k_\mathrm{p}\xi)H(-\xi). \tag{2.31}$$

$K_0(x)$ is the zeroth-order modified Bessel function of the second kind, which is exponentially decaying. As a result, the maximum longitudinal electric field resides at the axis (see Fig. 2.1b). In addition, we can infer that the radially nonuniform field will lead to energy spread increase of a beam with a finite radius.

The transverse field behind a point charge can be solved in a similar way from Eq. 2.27. But with the Panofsky-Wemzel theorem [10], it is easier to employ the relation between the longitudinal and transverse fields

$$\frac{\partial W_\parallel}{\partial r} = \frac{\partial W_\perp}{\partial z}. \tag{2.32}$$

This relation can also be directly obtained by doing r-derivative of Eq. 2.26 and z-derivative of Eq. 2.27. It follows that

$$G_{W_\perp}(r,\xi) = \int dz \frac{\partial G_{E_z}(r,\xi)}{\partial r} = 2ek_p^2 K_1(k_p r)\sin(k_p\xi)H(-\xi). \tag{2.33}$$

The longitudinal field produced by an arbitrary beam with the charge density $\rho_b = \rho_\parallel(\xi)\rho_\perp(r,\theta)$ is the convolution of the Green's function with the beam density

$$E_z(r,\theta,\xi) = 2k_p^2 \int_0^\infty \int_0^{2\pi} \int_{-\infty}^{\xi} \rho_b(r',\theta',\xi')r' K_0(k_p|r-r'|)\cos k_p(\xi-\xi')dr'd\theta'd\xi'. \tag{2.34}$$

Rewrite Eq. 2.34 as

$$E_z = \mathcal{Z}'(\xi)\mathcal{R}(r) \tag{2.35}$$

where

$$\mathcal{Z}'(\xi) = 4\pi \int_{-\infty}^{\xi} \rho_\parallel(\xi')\cos k_p(\xi-\xi')d\xi', \tag{2.36}$$

$$\mathcal{R}(r) = \frac{k_p^2}{2\pi}\int_0^\infty \int_0^{2\pi} \rho_\perp(r',\theta')r' K_0(k_p|r-r'|)dr'd\theta', \tag{2.37}$$

then we can quickly get

$$W_\perp = \mathcal{Z}(\xi)\mathcal{R}'(r), \tag{2.38}$$

where $\mathcal{Z}' = \frac{\partial \mathcal{Z}}{\partial \xi}$ and $\mathcal{R}' = \frac{\partial \mathcal{R}}{\partial r}$.

Figure 2.1b, c shows the longitudinal and transverse plasma wakefields sourced by a bi-Gaussian electron bunch. Apparently, the plasma density wave and the transverse wakefield share the same phase while the longitudinal electric field lags by a phase shift of $\pi/2$. It follows that the communally accelerating and focusing phase region suitable for a witness beam occupies $\pi/2$, i.e., one fourth wake period. Also a GeV/m-scale accelerating gradient is shown and it is inferred that, with strong transverse plasma wakefields, either the drive or the witness beam can be focused well during the propagation. The downside is the longitudinal field decreases with the radius, leading to an increase of the energy spread for a witness beam with a finite radius. In addition, the radial field varies nonlinearly along the radius, which is detrimental to the preservation of the witness beam emittance.

It is worth pointing out that, the perturbed plasma density and the resulting wake form due to a positron bunch follows exactly the same rule as in the electron driven case except for a phase difference of π. This is brought by the change of sign when it comes to positive charges. To be specific, the sign of the beam term at the right hand side of Eq. 2.11 changes to the positive. The plasma density perturbation in Eq. 2.17 thus changes its sign, which determines the sign change of the wakefields in Eqs. 2.34 and 2.38, as they depend on the plasma density according to Eqs. 2.26 and 2.27.

The above derivation of the linear theory has made two approximations. First, the beam density is essentially smaller than the plasma density, so that the plasma response to the beam can be treated perturbatively. Second, the beam is assumed ultra-relativistic travelling at the speed of light c, which suggests that the beam current remains in the time scale of ω_p^{-1} and the radial motion of the beam particles under the plasma fields is not considered. Setting $v = c$ obvisously simplifies solving the equations, yet a more general analysis involving the relativistic effect of the beam on plasma waves can be found in Keinigs and Jones's paper [11].

2.2.3 *Limit of Linear Theory*

In the previous section, we have derived the plasma fields in response to an arbitrary relativistic beam with a charge density form $\rho_b = \rho_\parallel(\xi)\rho_\perp(r)$. For a gaussian bunch, the corresponding longitudinal profile is $\rho_\parallel(\xi) = qn_b e^{-\xi^2/2\sigma_z^2}$ and the transverse form is $\rho_\perp(r) = e^{-r^2/2\sigma_r^2}$. Here q is the charge of particle ($+e$ for the positron bunch and $-e$ for the electron bunch). Substituting into Eqs. 2.35, 2.36 and 2.37, we have the on-axis longitudinal electric field after some arrangements

$$E_z(0,\xi) = \sqrt{2\pi}(q/e)(m_e c\omega_p/e)(n_b/n_p)(k_p\sigma_z e^{-k_p^2\sigma_z^2/2})R(0)\cos(k_p\xi), \quad (2.39)$$

where

$$R(0) = \left(\frac{k_p^2\sigma_r^2}{2}\right)(e^{k_p^2\sigma_r^2/2})\Gamma(0, k_p^2\sigma_r^2/2), \quad (2.40)$$

and $\Gamma(\alpha,\beta) = \int_\beta^\infty t^{\alpha-1}e^{-t}dt$.

Combining Eqs. 2.39 and 2.40, and rewriting the beam density in terms of the particle number $N_b = (2\pi)^{3/2}\sigma_r^2\sigma_z n_b$, we can get the wake amplitude in the following expression

$$E_{za} = qN_b k_p^2\left\{(e^{-k_p^2\sigma_z^2/2})(e^{k_p^2\sigma_r^2/2})\Gamma(0, k_p^2\sigma_r^2/2)\right\}. \quad (2.41)$$

It shows that given a beam with fixed N_b, σ_r, and σ_z, the maximum wake amplitude depends explicitly on k_p^2. The formula can be rewritten as

$$\begin{aligned} E_{za} &= \frac{qN_b}{\sigma_r\sigma_z}\left\{(k_p\sigma_z)(e^{-k_p^2\sigma_z^2/2})(k_p\sigma_r)(e^{k_p^2\sigma_r^2/2})\Gamma(0, k_p^2\sigma_r^2/2)\right\} \\ &= \frac{qN_b}{\sigma_r\sigma_z}\Omega(k_p\sigma_z, k_p\sigma_r), \end{aligned} \quad (2.42)$$

where $\Omega(k_p\sigma_z, k_p\sigma_r)$ can be treated as a function of $k_p\sigma_z$ and $r_a = \sigma_r/\sigma_z$, which is the beam's aspect ratio. It is found that for $\sigma_r/\sigma_z < 0.1$ (long and narrow beams), the optimal plasma wavenumber k_{p1} satisfies $k_{p1}\sigma_z = \sqrt{2}$ so that $\partial\Omega/\partial k_p\big|_{k_{p1}} = 0$ [3].

At the optimal plasma wavenumber k_{p1}, the maximum value of Ω depends only on r_a. It follows that the maximum wake amplitude can be expressed as

$$E_{zm} = \frac{qN_b}{\sigma_r\sigma_z}\Theta(\sigma_r/\sigma_z). \tag{2.43}$$

where $\Theta(\sigma_r/\sigma_z) = \Omega(k_{p1}\sigma_z, k_{p1}\sigma_r)$. Let $\Pi(\sigma_r/\sigma_z) = \sigma_z/\sigma_r\Theta(\sigma_r/\sigma_z)$, the formula becomes a slowly changing logarithmic function of r_a

$$E_{zm} = \frac{qN_b}{\sigma_z^2}\Pi(r_a). \tag{2.44}$$

However, the weak dependence on r_a is non-negligible as this term varies considerably for typical experimental parameters. In the limit of $\sigma_r/\sigma_z \ll 1$, an accurate asymptotic expansion for $\Pi(r_a) = 2/\mathrm{e}[-0.577 - 2\ln(r_a)]$ is obtainable. The maximum wake amplitude which occurs at $k_p\sigma_z = \sqrt{2}$ can be expressed as

$$E_{zm} = \frac{qN_b}{\sigma_z^2}\left\{\frac{4}{\mathrm{e}}[0.05797 - \ln(k_p\sigma_r)]\right\}, \tag{2.45}$$

where e is the exponential constant, or in a normalized form

$$\frac{eE_{zm}}{mc\omega_p} \approx 1.3\frac{q}{e}\frac{n_b}{n_p}k_p^2\sigma_r^2\ln\left(\frac{1}{k_p\sigma_r}\right), \tag{2.46}$$

where the italic e is the elementary charge and the constant term 0.05797 is neglected as $k_p\sigma_r \ll 1$. We can also rewrite the expression in a convenient engineering formula

$$E_{zm} \approx (236\,\mathrm{MV/m})\left(\frac{q}{e}\right)\left(\frac{N_b}{4\times 10^{10}}\right)\times\left(\frac{0.06\,\mathrm{cm}}{\sigma_z}\right)^2\ln\left(\sqrt{\frac{10^{16}\,\mathrm{cm}^{-3}}{n_p}}\frac{50\,\mathrm{mm}}{\sigma_r}\right). \tag{2.47}$$

The linear theory is valid if all the assumptions are satisfied. That is, the beam density is much smaller than the plasma density ($n_b/n_p \ll 1$), and the beam is ultra-relativistic. As a result, the fluid velocity is non-relativistic ($v \ll c$) and the normalized wake amplitude is less than unity, $eE_{zm}/mc\omega_p \ll 1$.

Given a fixed amount of beam charge, when shrinking the spot size till $n_b/n_p > 1$, the normalized electric field can still be less than unity as the term $n_b/n_pk_p^2\sigma_r^2$ in Eq. 2.46 dominates. It can get very large when keeping reducing the spot size due to divergence of the logarithmic term. In physics, it corresponds to the break down of the fluid theory. The plasma electrons are blown out radially by the very narrow beam, and their trajectories cross. In addition, the nonlinear effects come into play. For instance, the wake wavelength increases due to the change of the relativistic mass of plasma electrons.

By comparing with the simulation results, Ref. [3] demonstrates validity of the linear theory in the weakly nonlinear limit for the narrow beams. To be specific, for the electron driver, the peak accelerating field agrees with the Eq. 2.46 up to $n_b/n_p \approx 10$. For $n_b/n_p > 10$, it follows a different expression

$$\frac{eE_{zm}}{mc\omega_p} \approx 1.3\frac{q}{e}\frac{n_b}{n_p}k_p^2\sigma_r^2 \ln\left(\frac{1}{\sqrt{\Lambda/10}}\right), \tag{2.48}$$

where Λ denotes the normalized charge per unit length $\Lambda \equiv n_b/n_p k_p^2\sigma_r^2$. Basically the original logarithmic term $\ln(\frac{1}{k_p\sigma_r})$ is changed into $\ln(\frac{1}{\sqrt{\Lambda/10}})$. We can see now the wakefield depends on the charge of the beam instead of its density and reducing the beam radius has little effect as now the "blow-out" radius is greatly larger than the beam spot size.

The validity of the linear theory for the positron bunch is more strict. For $n_b/n_p > 1$, the normalized peak field changes to the form

$$\frac{eE_{zm}}{mc\omega_p} \approx 1.3\frac{q}{e}\frac{n_b}{n_p}k_p^2\sigma_r^2 \ln\left(\frac{1}{\sqrt{\Lambda}}\right), \tag{2.49}$$

This is because instead of expelling plasma electron away like the electron driver, the positron driver sucks them in. The trajectories cross inevitably when the plasma electrons reach the axis.

In the strongly nonlinear regime ($\Lambda \gg 1$), the above scaling laws for electron and positron drivers are no longer applicable. Simulations confirm that the wake amplitude is much smaller than that given by the above expressions.

2.3 Plasma Wakefield in Nonlinear Regime

In the previous section, we have elucidated the linear fluid theory in plasma wakefield excitation, where the driving beam density n_b is assumed substantially less than the plasma density n_p, so that the issue can be treated perturbatively. However, with intense drivers, the linear theory breaks down as more complicated and nonlinear effects come into play. For example, the plasma electron trajectories become not laminar but cross with each other. This brings in complexities and nonlinearities in describing the turbulent plasma. Also the relativistic mass effect gets involved as the plasma motion becomes relativistic. The fields are electromagnetic but can be treated by the full set of Maxwell's equations once the currents and charge densities are known. There have been studies regarding one-dimensional nonlinear fluid theory [12, 13] in the early date. Yet the break-through experiments [14–17] operating in the so-called "blow-out" regime [18] seek theoretical explanations in terms of multidimensional relativistic plasma wakefields.

The multidimensional nonlinear kinetic theory was developed by Lu et al. [4, 5] which turns out to be in good agreement with the simulations and hence applies to the experiments better. In Lu's theory, it assumes the plasma electrons within a "blow-out" radius are completely expelled by the space charge field of an intense electron driver or the ponderomotive force of a strong laser. The expelled plasma electrons form a thin dense sheath along the "blow-out' radius which is surrounded by weakly responding plasma electrons.

The transition from the linear to the nonlinear regime is determined by the plasma electron trajectories. In the linear regime, the maximum radius $r_{\mathrm{m}}(r_0)$ of the perturbed plasma electron always follows a monotonically increasing function of the initial radius r_0. Hence, the electron trajectories do not cross and the plasma is a cold laminar flow (see Fig. 2.2a). However, in the nonlinear regime, there exists an plasma electron initially in sufficiently small radius whose trajectory crosses with that of another electron. In other words, $r_{\mathrm{m}}(r_0)$ has local minima and maxima. As a result, it is possible to form a pure ion column below this small radius and the plasma electrons concentrate in a narrow sheath out of this ion column (see Fig. 2.2b). The "blow-out" region completely devoid of plasma electrons but filled with uniform plasma ions is also termed as a bubble [19]. Assuming the sheath width Δ is much less than the bubble radius r_{b}, it is feasible to get the trajectory of the bubble radius $r_{\mathrm{b}}(\xi)$, i.e., the innermost trajectory of the plasma electrons. This quantity further determines the wakes.

In the ultrarelativistic limit, where $n_{\mathrm{b}} \gg n_{\mathrm{p}}$ and $r_{\mathrm{m}} \gg 1$, the trajectory of a particle moving along the bubble can be expressed as a second-order differential equation

$$r_{\mathrm{b}}\frac{d^2 r_{\mathrm{b}}}{d\xi^2} + 2\left[\frac{dr_{\mathrm{b}}}{d\xi}\right]^2 + 1 = \frac{4I(\xi)}{r_{\mathrm{b}}^2}, \tag{2.50}$$

where $I(\xi) = \int_0^{r\gg\sigma_{\mathrm{r}}} r n_{\mathrm{b}} dr$ is the axial driving beam current. Note the use of natural units, i.e., $m_{\mathrm{e}} = c = e = 1$, and $\xi = ct - z$. It follows that the bubble radius can be directly obtained by integrating the Eq. 2.50. As the bunch length is typically much less than the bubble length, we can ignore the effect of the driving source on the majority of the trajectory or the bubble shape. After setting $\lambda(\xi)$ to zero, the equation of the bubble radius resembles a circle equation except the coefficient of the second term is 2 instead of 1. Near the top of the bubble, $dr_{\mathrm{b}}/d\xi \to 0$, so the trajectory $r_{\mathrm{b}}(\xi)$ draws a circle. But at the rear of the bubble, the additional term $[dr_{\mathrm{b}}/d\xi]^2$ bends the trajectory downward more quickly (see Fig. 2.2b).

The pseudo-potential for $r \le r_{\mathrm{b}}$ is given by

$$\psi = \frac{r_{\mathrm{b}}^2(\xi)}{4}(1+\alpha) - \frac{r^2}{4}, \tag{2.51}$$

where $\alpha = \Delta/r_{\mathrm{b}}$ and $\alpha \ll 1$. The potential is in the unit of $m_e c^2/e$. The length scales like r_{b} and r are in the unit of c/ω_{p}. The longitudinal wakefield E_z and the transverse field $W_\perp = E_{\mathrm{r}} - B_\theta$ in the ion channel then read

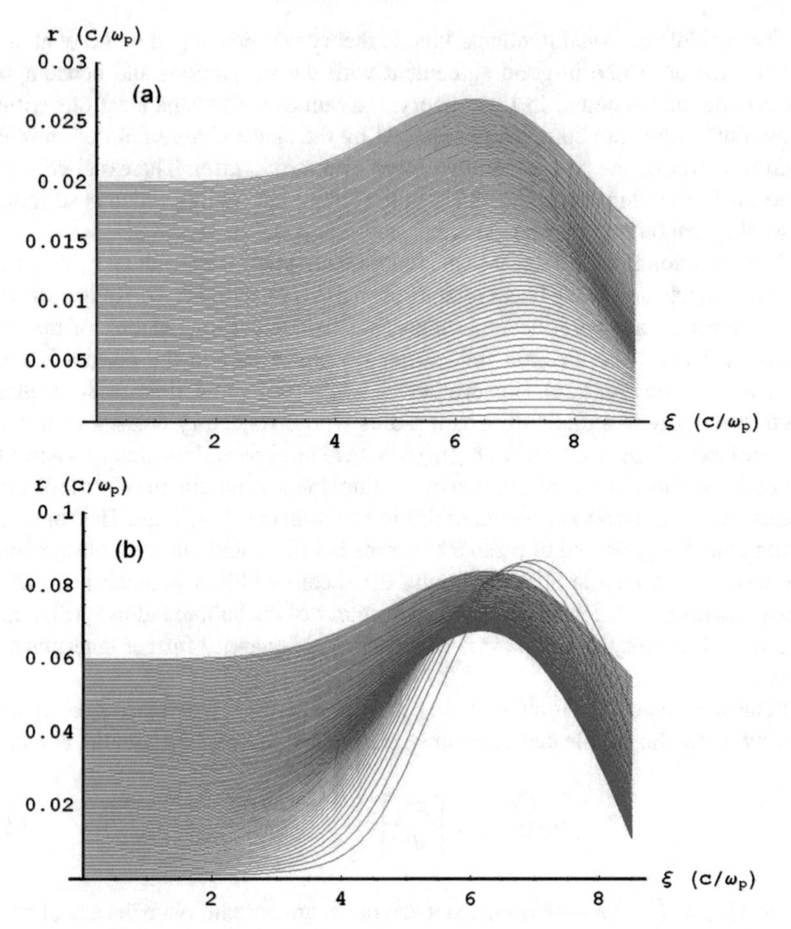

Fig. 2.2 Trajectories of the plasma electrons in the linear (***a***) and nonlinear (***b***) regime. The drive beam traveling to the left centres at $\xi = 5$. Image reproduced from Ref. [5]

$$E_z = \frac{d\psi}{d\xi} = \frac{1}{2} r_b(\xi) \frac{dr_b}{d\xi}, \tag{2.52}$$

$$W_\perp = -\frac{d\psi}{dr} = \frac{r}{2}, \tag{2.53}$$

which are normalized to the wave-breaking field $m_e c\omega_p/e$.

Equation 2.52 illustrates that the longitudinal field is constant along the radius as it is independent of r. Hence, there is no energy spread increase due to a finite radius of the beam. Equation 2.53 shows that the focusing is linear radially and constant along the longitudinal position, thus is conducive to the preservation of the beam emittance. We can conclude that the wakefields formed in the ion channel in the nonlinear regime are ideal for the accelerated beams and are significantly superior to that in the linear regime in terms of preserving the beam quality.

2.4 Beam Loading Characteristics

So far we have introduced how the wakefields are excited by driving sources in the linear and nonlinear regime. When a trailing beam is loaded into the accelerating wake phase behind the driver, it gets accelerated by the wakefield. In this way, the high energy is transferred from the driver to the witness bunch. Given a trailing beam with a finite length, the particles in different ξ-positions will see different accelerating gradients and thus gain energy in different rates. This leads to an increase of the beam energy spread. In theory, if the beam is ultra-short and with low charge, the growth of energy spread is negligible. Yet the energy extracted from the wake is low, so is the energy conversion efficiency. In this section, we demonstrate that by loading a properly tailored bunch to flatten the wakefield, the undesired increase of the energy spread can be eliminated. Following that we illustrate the transverse dynamics of the witness beam. Finally, we introduce the concepts of the transformer ratio and energy transfer efficiency. Note that in this section we specify the witness bunch as an electron bunch. The accelerating characteristics of a positron bunch is different and will be demonstrated exclusively in Chap. 5.

2.4.1 Beam Loading in the Linear Plasma Wakefield

In the linear regime, the total wakefield seen by the witness bunch is the superposition of the wakefield sourced by the driver and its self-wake, which reads [2]

$$E_{\rm t} = E_0 \cos k_{\rm p}\xi - 4\pi \int_{\xi_0}^{\xi} \rho_{\rm b}(\xi') \cos k_{\rm p}(\xi - \xi')d\xi', \tag{2.54}$$

where the second term is the self-field, $\rho_{\rm b}$ is the bunch density, $\xi = ct - z$, ξ_0 is the position of the head of the bunch. To lower the energy spread, $E_{\rm t}$ needs to be constant within the beam, i.e., $E_{\rm t} = E_0 \cos k_{\rm p}\xi_0$. S. Van der Meer [6] suggests that a finite bunch with a ramping down density allows for a flat field behind the bunch. Assuming $\rho_{\rm b}(\xi) = a\xi + b$, Ref. [2] resolves the coefficients a and b and gives

$$\rho_{\rm b}(\xi) = -\frac{k_{\rm p}E_0}{4\pi}[(k_{\rm p}\cos k_{\rm p}\xi_0)\xi + (\sin k_{\rm p}\xi_0 - k_{\rm p}\xi_0 \cos k_{\rm p}\xi_0)], \tag{2.55}$$

where $\frac{k_{\rm p}E_0}{4\pi e} = n_1$ is the density perturbation associated with the wave. This form indicates a trianglular bunch with the peak density at the head of the bunch or a trapezoidal shape if the back of the bunch is truncated earlier.

The maximum bunch length corresponds to the bunch density starting changing its sign, that is,

$$L_{max} = k_{\rm p}^{-1} \tan k_{\rm p}\xi_0. \tag{2.56}$$

The peak bunch density is the density at the bunch head $\xi = \xi_0$

$$\rho_b^{max} = -en_1 \sin k_p\xi_0. \tag{2.57}$$

As a result, the maximum number of particles being accelerated is

$$N = A\rho_b^{max} L_{max}/(-2e) = N_0 \frac{\sin^2 k_p\xi_0}{2\cos k_p\xi_0}, \tag{2.58}$$

where N_0 is $n_1 A/k_p$ and A is the cross-sectional area of the beam. The beam loading efficiency is

$$\eta_b = 1 - \frac{E_t^2}{E_0^2} = \sin^2 k_p\xi_0. \tag{2.59}$$

It denotes the fraction of wave energy absorbed by the witness bunch. We see there is a tradeoff between the accelerating gradient E_t and the efficiency η_b or the loaded particle number N. They all depend on the starting phase $k_p\xi_0$ of the beam loading. At the extreme case when $k_p\xi_0 = 0$, $E_t = E_0$ and the beam loading efficiency is zero, which is reasonable as Eqs. 2.56 and 2.58 imply zero witness bunch length and zero charge. The loading efficiency reaches 100% for $k_p\xi_0 = \pi/2$, whereas now the bunch is infinitely long and with infinite charge, which is impractical in the linear regime. As a compromise, when placing the beam front $\pi/3$ ahead of the peak field, the accelerating gradient reaches half the peak accelerating amplitude and the energy extraction is up to 75% without energy spread.

2.4.2 Beam Loading in the Plasma Bubble

In the nonlinear regime, although the total wake is not simply a superposition of the driven wake and the self-wake, the beam with a longitudinally trapezoidal current distribution turns out to be able to produce a flat field likewise [7, 8]. Assuming the beam current follows $I(\xi) = c\xi + d$ where c and d are coefficients and substitutes it into the right hand side of the Eq. 2.50, the beam profile can be calculated as

$$I(\xi) = -E_{t0}\xi + \sqrt{E_{t0}^4 + \frac{R_b^4}{2^4}} + E_{t0}\xi_0, \tag{2.60}$$

under the requirement that the wake is constant within the bunch, i.e., $E_z|_{\xi>\xi_0} = 1/2r_b(dr_b/d\xi)|_{\xi=\xi_0} = -E_{t0}$. Here E_{t0} is a positive value and R_b is the maximum bubble radius. The bubble shape within the bunch can be expressed as

$$r_b^2 = r_0^2 - 4E_{t0}(\xi - \xi_0), \tag{2.61}$$

where r_0 is the bubble radius at $\xi = \xi_0$ and r_b corresponds to the bubble radius at any position within the bunch. The maximum bunch length occurs when the bunch extends to the back of the bubble, i.e., $r_b = 0$. As a result, $L_{max} = \frac{r_0^2}{4E_{t0}}$ and the maximum total charge can be loaded is

$$Q_{total} = 2\pi L_{max} \frac{I(\xi_0) + I(\xi_0 + L_{max})}{2} = \frac{\pi}{16} \frac{R_b^4}{E_{t0}}. \tag{2.62}$$

The total energy absorption per unit length for the trapezoidal bunch is then

$$Q_{total} E_{t0} = \frac{\pi R_b^4}{16}. \tag{2.63}$$

It indicates that the energy extraction is independent of the accelerating gradient. Therefore, despite a tradeoff between the total loading charge and the accelerating gradient, the beam loading efficiency can be made close to 100%.

2.4.3 Transverse Characteristics of the Loaded Beam in the Nonlinear Regime

In vacuum, despite the cancellation of the space charge field and the self-magnetic field, the beam will inevitably diverge driven by its initial angular spread. Fortunately, in the plasma wakefield acceleration, the plasma provides strong transverse focusing to the beam so that it can either keep driving the wakefields as a driver or be accelerated as a witness bunch. In the linear regime, the focusing field is nonlinear along the radius according to Eq. 2.38. Hence, the particles perform anharmonic transverse oscillations, leading a large growth of the final bunch emittance.

On the contrary, in the nonlinear regime, the witness electron beam resides in a pure ion column with a uniform ion density. The focusing field is linear with radius (Eq. 2.53) and reads in the SI unit

$$E_r = \frac{1}{2} \frac{n_0 e}{\epsilon_0} r, \tag{2.64}$$

where the ion density is equal to the initial plasma electron density n_0. The motion of an electron in the ion column follows

$$\gamma m_e \frac{d^2 r}{dt^2} = -e E_r, \tag{2.65}$$

which turns into the harmonic equation with respect to z

$$\gamma m_e c^2 \frac{d^2 r}{dz^2} = -\frac{1}{2}\frac{n_0 e^2}{\epsilon_0} r. \tag{2.66}$$

After simplification it becomes

$$\frac{d^2 r}{dz^2} + k_\beta^2 r = 0, \tag{2.67}$$

where $k_\beta^2 = k_p^2/2\gamma$ and k_p is the plasma wavenumber. $\beta = 1/k_\beta$ is commonly named as the betatron function. A general solution to this equation is

$$r(z) = r_0 e^{ik_\beta z}, \tag{2.68}$$

which indicates a transversely harmonic oscillation of the particle at the frequency $\omega_\beta = \omega_p/\sqrt{2\gamma}$.

As to a bunch of charged particles, its envelope evolution under the initial emittance can be described with an equation [20]

$$\frac{d^2\sigma_r(z)}{dz^2} + k_\beta^2 \sigma_r(z) = \frac{\varepsilon^2}{\sigma_r^3(z)}, \tag{2.69}$$

which is similar to that for individual particles. Here σ_r is the bunch transverse RMS size and ε is the geometrical beam emittance. It follows that if σ_r is large, the term $k_\beta^2\sigma_r$ dominates the term ε^2/σ_r^3 and σ_r'' becomes negative, then the beam shrinks due to the external force. Similarly, the beam expands due to its emittance when σ_r is small. As a result, the beam matches with the focusing structure when $\sigma_r'' = \sigma_r' = 0$, that is,

$$k_\beta^2 \sigma_r = \frac{\varepsilon^2}{\sigma_r^3}. \tag{2.70}$$

Thus, the equilibrium radius of the beam can be written as

$$\sigma_r = \sqrt{\beta\varepsilon}. \tag{2.71}$$

The beam emittance is an important property of a beam in accelerator physics. It denotes the area of the bunch phase space (x, x') and can be expressed as

$$\varepsilon = \sqrt{\langle x^2\rangle\langle x'^2\rangle - \langle xx'\rangle^2}, \tag{2.72}$$

where x is the radial position of each individual particle and $x' = \mathrm{d}x/\mathrm{d}z = p_x/p_z$ is its angular divergence, $p_z \approx p$ is the longitudinal momentum. The angle brackets $\langle\rangle$ do the average among all particles. Since the beam is normally under acceleration, it is more convenient to use the normalized emittance

$$\varepsilon_n = \sqrt{\langle x^2\rangle\langle p_x{}^2\rangle - \langle xp_x\rangle^2}, \tag{2.73}$$

where the transverse momentum p_x replaces x' because it decreases with the increasing p during acceleration. ε_n can be approximated by $\langle\gamma\rangle\,\varepsilon$. If a beam is injected into the plasma with an equilibrium radius, its normalized emittance will be unchanged. It follows from Eq. 2.71 that σ_r is proportional to $\gamma^{-1/4}$. It indicates a weak dependence of the equilibrium beam radius on the particle energy, which may allow for adiabatic matching of the bunch size to the plasma focusing upon acceleration.

It is worth noting that while the bunch envelope does not oscillate in the matched case, the particles do. The particles oscillate with the betatron wavelength $\lambda_\beta = \sqrt{2\gamma}\lambda_p$, which is much longer than the plasma wavelength. This concept had been used in the quasi-static approximation in the particle-in-cell code where the beam is considered rigid while calculating the plasma response. It will be introduced in Sect. 2.5.

2.4.4 *Transformer Ratio and Efficiency*

The transformer ratio R is an important property of the plasma wakefield accelerators to evaluate the energy transfer from the driver to the witness bunch. It is a ratio of the maximum accelerating field amplitude behind the driver and the maximum decelerating field amplitude seen by the driver: $R = E_+/E_-$ [1]. It can be calculated by assuming two rigid bunches with zero length, i.e., the driver with N_1 particles and energy W_1 per particle and the witness bunch with N_2 particles and energy W_2 per particle. Given $E(\xi)$ is the longitudinal decelerating field excited by a unit charge at the distance ξ behind it, the energy loss rate by the driver located at $\xi_1 = 0$ is

$$\frac{d(N_1W_1)}{dz} = -(N_1e)N_1eE(0) = -N_1^2e^2E(0). \tag{2.74}$$

The witness bunch located at ξ_2 behind the driving bunch sees the field induced by the driver and its own wakefield, thus its energy change rate is

$$\frac{d(N_2W_2)}{dz} = -(N_2e)[N_2eE(0) + N_1eE(\xi_2)] = -N_2^2e^2E(0) - N_1N_2e^2E(\xi_2). \tag{2.75}$$

Note that N_1 and N_2 are constant. The energy change rate is independent of the sign of the charge e but it depends on the sign of the $E(\xi_2)$, that is, the phase of the witness bunch in the wakefield of the driver.

The total energy change must not exceed zero due to energy conservation, i.e.,

$$\frac{d(N_1W_1)}{dz} + \frac{d(N_2W_2)}{dz} = -(N_1^2 + N_2^2)e^2E(0) - N_1N_2e^2E(\xi_2) \leq 0. \tag{2.76}$$

This inequality equation must hold for arbitrary N_1 and N_2, then we get

$$-E(\xi_2) \le 2E(0). \tag{2.77}$$

As a result, the accelerating field gradient G seen by a particle in the witness bunch is

$$G = \frac{d(W_2)}{dz} \le (2N_1 - N_2)e^2 E(0). \tag{2.78}$$

The maximum energy gained by the witness bunch occurs when the driver loses all its energy and the corresponding deceleration distance is

$$L = \frac{W_1}{N_1 e^2 E(0)}. \tag{2.79}$$

The maximum energy gain by the witness bunch is hence

$$\Delta W_2 = GL \le W_1 \left(2 - \frac{N_2}{N_1}\right). \tag{2.80}$$

When $N_2 = 0$, $\Delta W_2 \le 2W_1$, which implies the basic limitation of a co-linear wakefield accelerator, that is, the total energy gain per particle in a witness bunch cannot exceed twice the initial energy per particle in the driving bunch. This suggests possible energy transfer from a high-charge, low energy bunch to a low-charge bunch via plasma wakefields, which is constructive in a staged accelerator. Therein the witness bunch keeps extracting energy from the low energy bunch in each plasma stage and eventually reaches very high energy.

The transformer ratio reads

$$R = \frac{|dW_2/dz|}{|dW_1/dz|} = \frac{\Delta W_2}{W_1} \le \left(2 - \frac{N_2}{N_1}\right). \tag{2.81}$$

Regardless of the assumption of infinitely short bunches made in the above derivation, it can be shown that the limit on the transformer ratio applies to any symmetric bunch distribution. For an asymmetric bunch where the bunch current profile is tailored longitudinally or by using multiple driving bunches, the transformer ratio can greatly surpass 2 [1, 2, 21, 22]. The transformer ratio has been measured in the plasma wakefield experiments [23–25].

The energy transfer efficiency from the driver to the witness bunch is easily obtained:

$$\eta = \frac{N_2 \Delta W_2}{N_1 W_1} = \frac{N_2}{N_1}\left(2 - \frac{N_2}{N_1}\right). \tag{2.82}$$

The maximum efficiency occurs when $N_1 = N_2$ and all the driving energy transfers to the witness bunch, but it limits the transformer ratio to 1.

2.5 Particle-in-Cell Simulations

Experiments which take into account the uncertainties, imperfections and variations in reality are indispensable. Experimental results are generally powerful and persuasive proofs whether the studied case works in the real world. Regardless, numerical simulations play an essential role in studying the plasma wakefield acceleration before the corresponding experiments are conducted. This is because the underlying physics generally involves nonlinear effects and relativistic plasma and beams. It requires proper numerical modelling tools to explore the complicated beam and plasma dynamics. In addition, the simulations offer greater flexibility and quicker operation in adjusting the parameters to basically any values compared to experiments, which can only vary parameters in a small range limited by the engineering technologies. Although for some modelling cases it requires high performance computing resources, the large scale clusters are more available than the particle accelerators with required parameters. Overall, simulation tools can save scientists substantial time and costs and also get them to understand the underlying physics under parameters either obtainable in the experiments or far beyond the current technologies.

In this section, we will introduce the algorithm of the particle-in-cell (PIC) method, which is the basis of the full relativistic PIC code. Following that we elucidate the quasi-static PIC code which acts as the main numerical tool in our research.

2.5.1 *Full Relativistic PIC Codes and the Algorithm*

The full electromagnetic relativistic PIC code has been proved reliable in describing the kinetic properties of plasma. It solves the Maxwell equations together with the motion equation of the particles [26]. To formulate the problem, let's rewrite the Maxwell's equations in SI units:

$$\epsilon_0 \frac{\partial \boldsymbol{E}}{\partial t} = \frac{1}{\mu_0} \nabla \times \boldsymbol{B} - \boldsymbol{J}, \tag{2.83}$$

$$\frac{\partial \boldsymbol{B}}{\partial t} = -\nabla \times \boldsymbol{E}, \tag{2.84}$$

$$\nabla \cdot \boldsymbol{E} = \frac{\rho}{\epsilon_0}, \tag{2.85}$$

$$\nabla \cdot \boldsymbol{B} = 0, \tag{2.86}$$

where ϵ_0, μ_0 are the vacuum permittivity and permeability respectively. We see from Eqs. 2.83 and 2.84 that the electric and magnetic fields $\boldsymbol{E}, \boldsymbol{B}$ evolve with time. The source term is the current density $\boldsymbol{J}$, which is produced by the charge motion in the system. By taking the divergence of Eq. 2.83 and the charge continuity equation

$$\frac{\partial \rho}{\partial t} = -\nabla \cdot \boldsymbol{J}, \tag{2.87}$$

we can get the Poisson Eq. 2.85 true. As there is no magnetic charge, Eq. 2.86 is always valid. Hence, we can simplify the problem to solving the first two Maxwell's equations while considering the last two as the initial conditions, provided that the charge and current densities always satisfy the continuity equation.

In physics, the electromagnetic fields are continuous in space and time and any real plasma contains an extremely large number of charged particles. It is hardly possible to store the positions and momenta of all individual charged particles, as the required memory is far beyond the capability of computers available. PIC codes take some approximations and employ a significantly smaller number of numerical macro-particles to represent the real particles. Each numerical macro-particle substitutes a cloud of many real particles which occupy a finite volume in space and move together with the same velocity [27]. The macro-particle carries the combined charge and mass of the real particle clump. In addition, the fields are discretized in simulations. A finite-difference time-domain algorithm involving Maxwell's equations in isotropic media was developed by Yee [28]. The Yee lattice is popularly adopted where the electric field components are defined in the middle of the grid edges while the magnetic field components are in the centres of the cell faces. The code only calculates the field components in the grids. The components at other points if needed will be calculated by interpolation.

At the boundaries of the finite simulation window (domain), the grids are truncated, which will cause some unphysical reflections of the electromagnetic waves. Therefore, specific boundary conditions are set for the sake of accurate calculation of the fields at the boundaries. There are different types of boundary conditions to accommodate different situations, such as the periodic boundary condition where the fields and/or particles reaching one edge of the domain are wrapped round to the opposite boundary, and the perfect conducting condition where the incoming waves are dumped in a specified region at the simulation boundary and the particles once reaching the boundaries are removed from the simulation. The latter is usually used for the boundaries transverse to the propagation direction. Another type of longitudinal boundary condition allows the electromagnetic wave sources (like lasers) to be attached to a boundary (i.e., transmitted with as little reflection as possible) or the particles to be fully transmitted. For the simulations illustrated in Chap. 6, the left boundary of the window traveling to the right is set to fully transmit the proton bunch driver, while the right and transverse boundaries all follow the perfect conducting condition.

In PIC simulation, when calculating, the derivative is replaced by the finite difference. Define the start time of every time step ΔT as t_{n} and the end time as $t_{\mathrm{n+1}}$, the update of the electromagnetic fields is as follows:

(1) Magnetic field update by half time step

$$\boldsymbol{B}^{\mathrm{n}+\frac{1}{2}} = \boldsymbol{B}^{\mathrm{n}} - \Delta T/2(\nabla \times \boldsymbol{E}^{\mathrm{n}}), \tag{2.88}$$

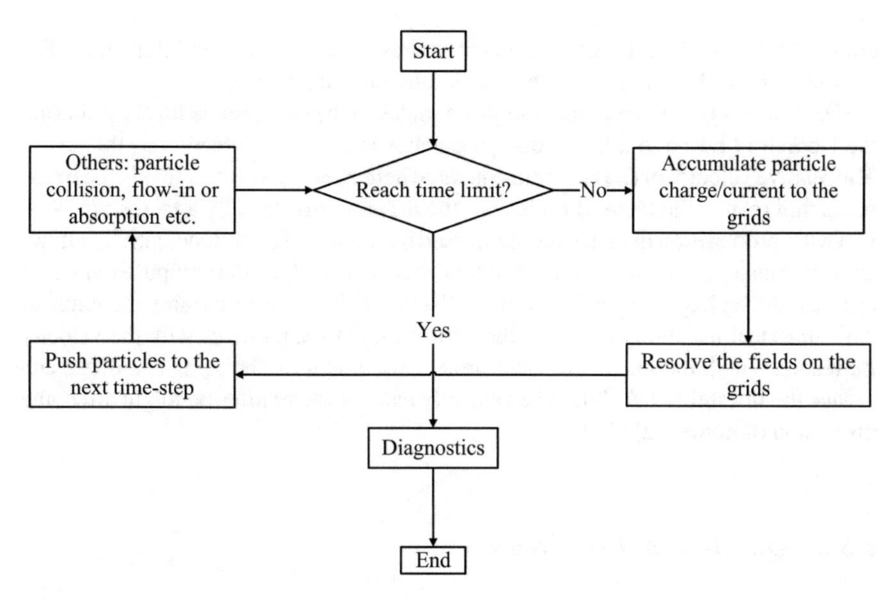

Fig. 2.3 Flowchart of the particle-in-cell code algorithm

(2) Electric field update by one time step

$$\epsilon_0 \boldsymbol{E}^{\mathrm{n}+1} = \epsilon_0 \boldsymbol{E}^{\mathrm{n}} + \Delta T \left(\nabla \times \frac{\boldsymbol{B}^{\mathrm{n}+\frac{1}{2}}}{\mu_0} - J \right), \tag{2.89}$$

(3) Magnetic field update by half time step

$$\boldsymbol{B}^{\mathrm{n}+1} = \boldsymbol{B}^{\mathrm{n}+\frac{1}{2}} - \Delta T/2 \left(\nabla \times \boldsymbol{E}^{\mathrm{n}+1} \right), \tag{2.90}$$

Once the electric and magnetic fields have been calculated, the update of the particle positions and momenta occurs according to the electromagnetic forces. The new particle distribution is used to calculate the current density, which is then used to calculate the fields. The flowchart in Fig. 2.3 shows a simplified cycle of the PIC code algorithm which is started after the initialization, i.e., setting macro-particles to represent the real particles, dividing the space into grids, and defining the electromagnetic fields on the grids.

Currently there exist many full PIC codes available for beam-plasma simulatons such as EPOCH [29], VSim [30], OSIRIS [31], WarpX [32], etc. While they represent the most fundamental model, they are computationally expensive especially in the high dimension and for a long acceleration (plasma) length. Plasma-based accelerators have multiple and disparate scales. The smallest one is the plasma wavelength (or the laser wavelength if a laser is involved). The medium one is the beam length which is usually comparable to the plasma wavelength or sometimes much larger when the self-modulated wakefields are under concern. The largest one is the acceleration distance which ranges from centimetres to meters or even kilometres. The enormous scale discrepancy makes the simulations rather heavy.

The Lorentz boost technique brings the scales to the comparable level by altering the laboratory frame to a frame that propagates in the same direction as the driver. The relative velocity of the new frame is βc, where $\beta = \sqrt{1-\gamma^{-2}}$ and γ is the transformation relativistic factor. In this way, the driver is stretched by a factor $\gamma(1+\beta)$ while the propagation distance is compressed by the same factor. It potentially allows an increased longitudinal grid size and time step. In principle, the computational cost can be reduced hugely by a factor of γ^2. The issue is in the new frame the unphysical numerical instability occurs if the streaming plasma particles with the velocity βc resonate with the numerical electromagnetic modes on the grid. The filters can reduce the instability but should be carefully used as heavy filtering might affect the dispersion of numerical modes.

2.5.2 *Quasi-static PIC Codes*

Given the fact that the Lorentz boost frame makes the simulations more complicated, the quasi-static approximation has been introduced into the PIC code which separates different scales analytically. In Sect. 2.4.3, we have demonstrated that the betatron oscillation wavelength of the particles relates to the plasma wavelength by $\lambda_\beta = \sqrt{2\gamma}\lambda_\mathrm{p}$. Therefore, for a relativistic beam, the time scale of its transverse evolution is much greater than the plasma wave period. This validates the quasi-static approximation, where it assumes the beam is rigid when calculating the plasma response.

In quasi-static PIC code, the fields are associated with the combined coordinate $\xi = z - ct$ instead of z and t. All derivatives with respect to the slow time t are now rewritten in terms of ξ. The fields are calculated from the front to the end of the driving beam at a specific time t. After the fields and the particle distribution are obtained, we can advance the driver by a large time step in t, as long as the time step is enough to resolve the shortest timescale of transverse beam oscillations. Note that the quasi-static code does not consider the radiation but only the static electromagnetic fields. The derivation of the quasi-static field equations is as follows.

We firstly rewrite the Maxwell's equations in Gaussian units in terms of the new combined variable ξ while neglecting the time t:

$$\nabla \times \boldsymbol{B} = \frac{4\pi}{c}\boldsymbol{J} - \frac{\partial \boldsymbol{E}}{\partial \xi}, \tag{2.91}$$

$$\nabla \times \boldsymbol{E} = \frac{\partial \boldsymbol{B}}{\partial \xi}, \tag{2.92}$$

$$\nabla \cdot \boldsymbol{E} = 4\pi\rho, \tag{2.93}$$

$$\nabla \cdot \boldsymbol{B} = 0. \tag{2.94}$$

Given that $\nabla \times \nabla \times \boldsymbol{B} = \nabla(\nabla \cdot \boldsymbol{B}) - \nabla^2 \boldsymbol{B} = -\nabla^2 \boldsymbol{B}$, after combining the curl of Eq. 2.91 and ξ-derivative of Eq. 2.92, we get

$$\nabla_\perp^2 \boldsymbol{B} = -\frac{4\pi}{c} \nabla \times \boldsymbol{J}, \tag{2.95}$$

where $\nabla^2 = \nabla_\perp^2 + \frac{\partial^2}{\partial \xi^2}$. Similarly, if we take the gradient of Eq. 2.93 and use the form $\nabla(\nabla \cdot \boldsymbol{E}) = \nabla \times \nabla \times \boldsymbol{E} + \nabla^2 \boldsymbol{E}$, we can obtain the transverse electric field components satisfying the equation

$$\nabla_\perp^2 \boldsymbol{E}_\perp = 4\pi \left(\nabla_\perp \rho - \frac{1}{c} \frac{\partial}{\partial \xi} \boldsymbol{J}_\perp \right), \tag{2.96}$$

and the longitudinal component

$$\nabla_\perp^2 E_\parallel = 4\pi \frac{\partial}{\partial \xi} \left(\rho - \frac{1}{c} J_\parallel \right) = \frac{4\pi}{c} \nabla_\perp \cdot \boldsymbol{J}_\perp, \tag{2.97}$$

where the continuity equation below has been used

$$c \frac{\partial}{\partial \xi} \rho = \frac{\partial}{\partial \xi} J_\parallel + \nabla_\perp \cdot \boldsymbol{J}_\perp. \tag{2.98}$$

The quasi-static PIC code works in the following procedure. First of all, it gathers the charge and current densities generated by the beam driver on the numerical grids. They contribute to the source terms of the field Eqs. 2.95–2.97. Then, the layer of macro-particles for plasma particles is seeded at the front boundary of the simulation window and advanced layer-by-layer towards the rear boundary. After the plasma particles pass the entire simulation domain, the fields and the particle density are determined on the grids. The calculated fields are then be used to modify the beam and advance it in time t by solving the motion equation for particles. We see it explicitly separates the fast scale of the driver and the slow scale of the acceleration [33].

In our research, we have been mostly relying on the quasi-static code as our studied cases are associated with long distances up to hundreds of meters but the gird must be fine enough (μm scale) to resolve the small beams. It is nearly impossible to simulate with full PIC codes on the modern machines. LCODE developed by Lotov et al. [34, 35] is our main choice, which is written based on the aforementioned quasi-static field equations. It is a 2D (2d3v) code with both plan and cylindrical geometries available. In simulation, the transverse boundary conditions correspond to a perfectly conducting tube. The unperturbed plasma with zero fields are taken as the longitudinal condition of the right boundary, because the simulation window moves at the speed of light to the right and nothing within the simulation window can reach this boundary. It takes any set of fields and velocities reaching there as input in calculating the plasma response afterwards. For the left boundary, it can fully transmit the incoming fields and particles. Note that all quantities in this code are dimensionless. The units of measure depend on the initial plasma density. For example, the times are in the units of ω_p^{-1}, the lengths are in the units of c/ω_p and the fields are in the units of $m_e c\omega_p/e$. The simulation results calculated with LCODE shown in the following chapters have been interpreted in SI units.

There are 3D quasi-static codes available like Quick-PIC [36] and VLPL [26], which are useful to see the 3D effects yet not suitable for our long-term simulations owing to huge computing resources required.

2.6 Summary

In this chapter, we have demonstrated the linear and nonlinear theories of plasma wakefield acceleration. In the linear regime where the beam density is essentially smaller than the plasma density, the plasma response is a perturbative quantity. The plasma fields oscillate sinusoidally at the plasma frequency in time. The transverse field is not linear to the radius and the longitudinal field is non-uniform along the radius. The linear theory breaks down when the beam density increases and the plasma trajectories cross. In the nonlinear regime where a plasma bubble (a pure ion column) is formed, the plasma fields are ideal for the electron acceleration. To be specific, the transverse plasma field is linear radially, which preserves the beam emittance. The longitudinal electric field is uniform in the radial direction, which causes no growth of the energy spread. We have also addressed the beam loading characteristics and associated properties like energy spread, beam normalized emittance, the transformer ratio and energy transfer efficiency. In the last section, we have introduced the PIC code algorithm which is devoted to resolving the Maxwell's equation and the particle motion equation. The quasi-static code which assumes the beam rigid while calculating the plasma response significantly advances the computing efficiency, and thus is superior in conducting heavy simulations.

References

1. Ruth RD, Morton P, Wilson PB, Chao A (1984) A plasma wake field accelerator. Part Accel 17(SLAC-PUB-3374):171
2. Katsouleas TC, Wilks S, Chen P, Dawson JM, Su JJ (1987) Beam loading in plasma accelerators. Part Accel 22:81–99, 1987
3. Lu W, Huang C, Zhou M, Mori W, Katsouleas T (2005) Limits of linear plasma wakefield theory for electron or positron beams. Phys Plasmas 12(6):063101
4. Lu W, Huang C, Zhou M, Tzoufras M, Tsung F, Mori W, Katsouleas T (2006) A nonlinear theory for multidimensional relativistic plasma wave wake fields. Phys Plasmas 13(5):056709
5. Lu W, Huang C, Zhou M, Mori W, Katsouleas T (2006) Nonlinear theory for relativistic plasma wakefields in the blowout regime. Phys Rev Lett 96(16):165002
6. van der Meer S (1985) Improving the power eficiency of the plasma wake field accelerator. Technical Report CM-P00058040

7. Tzoufras M, Lu W, Tsung F, Huang C, Mori W, Katsouleas T, Vieira J, Fonseca R, Silva L (2008) Beam loading in the nonlinear regime of plasmabased acceleration. Phys Rev Lett 101(14):145002
8. Tzoufras M, Lu W, Tsung F, Huang C, Mori W, Katsouleas T, Vieira J, Fonseca R, Silva L (2009) Beam loading by electrons in nonlinear plasma wakes. Phys Plasmas 16(5):056705
9. Polyanin AD, Nazaikinskii VE (2015) Handbook of linear partial diffierential equations for engineers and scientists. Chapman and hall/CRC, Boca Raton
10. Panofsky W, Wenzel W (1956) Some considerations concerning the transverse deflection of charged particles in radio-frequency fields. Rev Sci Instrum 27(11):967–967
11. Keinigs R, Jones ME (1987) Two-dimensional dynamics of the plasma wake field accelerator. Phys Fluids 30(1):252–263
12. Akhiezer AI, Polovin R (1956) Theory of wave motion of an electron plasma. Soviet Phys JETP 3:696–705
13. Dawson JM (1959) Nonlinear electron oscillations in a cold plasma. Phys Rev 113(2):383
14. Mangles SP, Murphy C, Najmudin Z, Thomas AGR, Collier J, Dangor AE, Divall E, Foster P, Gallacher J, Hooker C et al (2004) Monoenergetic beams of relativistic electrons from intense laser-plasma interactions. Nature 431(7008):535
15. Geddes C, Toth C, Van Tilborg J, Esarey E, Schroeder C, Bruhwiler D, Nieter C, Cary J, Leemans W (2004) High-quality electron beams from a laser wakefield accelerator using plasma channel guiding. Nature 431(7008):538
16. Faure J, Glinec Y, Pukhov A, Kiselev S, Gordienko S, Lefebvre E, Rousseau J-P, Burgy F, Malka V (2004) A laser-plasma accelerator producing monoenergetic electron beams. Nature 431(7008):541
17. Hogan M, Barnes C, Clayton C, Decker F, Deng S, Emma P, Huang C, Iverson R, Johnson D, Joshi C et al (2005) Multi-GeV energy gain in a plasma-wake field accelerator. Phys Rev Lett 95(5):054802
18. Rosenzweig J (1987) Nonlinear plasma dynamics in the plasma wakefield accelerator. IEEE Trans Plasma Sci 15(2):186–191
19. Pukhov A, Meyer-ter Vehn J (2002) Laser wake field acceleration: the highly non-linear broken wave regime. Appl Phys B 74(4–5):355–361
20. Muggli P (2017) Beam-driven, plasma-based particle accelerators. arXiv:1705.10537
21. Lotov K (2005) Efficient operating mode of the plasma wake field accelerator. Phys Plasmas 12(5):053105
22. Farmer J, Martorelli R, Pukhov A (2015) Transformer ratio saturation in a beam-driven wakefield accelerator. Phys Plasmas 22(12):123113
23. Blumenfeld I, Clayton C, Decker F, Hogan M, Huang C, Ischebeck R, Iverson R, Joshi C, Katsouleas T, Kirby N et al (2010) Scaling of the longitudinal electric field and transformer ratio in a nonlinear plasma wake field accelerator. Phys Rev ST Accel Beams 13(11):111301
24. Loisch G, Boonpornprasert P, Brinkmann R, Good J, Grofi M, Gruner F, Huck H, Krasilnikov M, Lishilin O, Martinez de la Ossa A et al (2018) Optimisation of high transformer ratio plasma wakefield acceleration at PITZ. In: Proceedings of IPAC18, Vancouver, BC, Canada
25. Loisch G, Asova G, Boonpornprasert P, Brinkmann R, Chen Y, Engel J, Good J, Gross M, Gruner F, Huck H et al (2018) Observation of high transformer ratio plasma wakefield acceleration. Phys Rev Lett 121(6):064801
26. Pukhov A (2015) Particle-in-cell codes for plasma-based particle acceleration. arXiv:1510.01071
27. Hockney RW, Eastwood JW (1988) Computer simulation using particles. CRC Press, Boca Raton
28. Yee K (1966) Numerical solution of initial boundary value problems involving maxwell's equations in isotropic media. IEICE Trans Antennas Propag 14(3):302307
29. Arber T, Bennett K, Brady C, Lawrence-Douglas A, Ramsay M, Sircombe N, Gillies P, Evans R, Schmitz H, Bell A et al (2015) Contemporary particle-in-cell approach to laser-plasma modelling. Plasma Phys Control Fusion 57(11):113001

30. Nieter C, Cary JR (2004) Vorpal: a versatile plasma simulation code. J Comput Phys 196(2):448–473
31. Fonseca RA, Silva LO, Tsung FS, Decyk VK, Lu W, Ren C, Mori WB, Deng S, Lee S, Katsouleas T et al (2002) Osiris: a three-dimensional, fully relativistic particle in cell code for modeling plasma based accelerators. In: International conference on computational science, Springer, Berlin, pp 342–351
32. Vay J-L, Almgren A, Bell J, Ge L, Grote D, Hogan M, Kononenko O, Lehe R, Myers A, Ng C et al (2018) Warp-x: a new exascale computing platform for beam plasma simulations. Nucl Instrum Methods Phys Res Sect A
33. Mora P, Antonsen TM Jr (1997) Kinetic modeling of intense, short laser pulses propagating intenuous plasmas. Phys Plasmas 4(1):217–229
34. Lotov K (2003) Fine wake field structure in the blowout regime of plasma wake-field accelerators. Phys Rev ST Accel Beams 6(6):061301
35. Sosedkin A, Lotov K (2016) LCODE: a parallel quasistatic code for computationally heavy problems of plasma wakefield acceleration. Nucl Instrum Methods Phys Res Sect A 829:350–352
36. Huang C, Decyk VK, Ren C, Zhou M, Lu W, Mori WB, Cooley JH, Antonsen TM Jr, Katsouleas T (2006) Quickpic: a highly efficient particle-in-cell code for modeling wake field acceleration in plasmas. J Comput Phys 217(2):658–679

Chapter 3
High Quality Electron Bunch Generation in a Single Proton Bunch Driven Hollow Plasma Wakefield Accelerator

3.1 Introduction

The storyline of this chapter is clearly seen from the title. We basically address the following questions one by one: why do we employ proton bunches as the drivers in the plasma wakefield acceleration? Why is a hollow plasma channel necessary and what features are hollow channel-based wakefields? How is the witness electron bunch accelerated and preserved with high beam quality?

To be specific, in the main simulation part, we introduce the basic concept of the proposed scheme and determine parameters for the PIC simulations. Then we elucidate the transverse dynamics of the driving and witness bunches in the hollow plasma and the longitudinal acceleration of the witness bunch, whose normalized emittance is shown preserved. We discuss the dependence of the energy gain, energy spread, normalized emittance and particle survival rate on the channel radius and the plasma density, and promote some approaches to further reduce the energy spread. Finally, we comment the validity of conducted simulations.

3.2 Considerations of Proton Bunches as Drivers

3.2.1 Competitive Energy Contents

In the past few decades, there has been tremendous progress in plasma-based wakefield accelerators. The electron beam energy has been advanced to 4.2 GeV driven by a 300 TW laser [1]. The energy gain in the beam-driven case is more impressive. For instance, SLAC in 2007 reported an energy doubling of the rear part of a driving electron bunch, which ended up with a witness beam energy as high as 84 GeV [2]. However, this energy record hasn't been broken since then. This is because, the

Y. Li, *Studies of Proton Driven Plasma Wakefield Acceleration*, Springer Theses,
https://doi.org/10.1007/978-3-030-50116-7_3

Table 3.1 Parameters for energetic bunches produced in different accelerator facilities

Parameters	SLAC-FFTB	ILC	PS	SPS	LHC
Bunch population, 10^{10}	1.8	2	13	11.5	11.5
Particle energy, GeV	42	250	24	450	7000
Total energy, kJ/bunch	0.12	0.8	0.5	8.3	129

obtainable energy by the witness particles in one acceleration stage is physically hindered by the energy content of the driver owing to the limited transformer ratio, and the energy of the driving electron beam used here has reached the limit that current accelerators can produce. Multi-staging acceleration can, in principle, boost the particle energy stage by stage to the energy frontier. Nonetheless, it is technically challenging as it requires tight synchronization and alignment of all the drivers and the witness bunch and of all accelerator modules.

Under this circumstance, proton bunches stand out due to existing huge energies and large populations. As seen from Table 3.1, the proton bunch energy produced in the LHC is orders of magnitude larger than the highest electron bunch energy produced at the SLAC. Even the 40 year-old Super Proton Synchrotron (SPS) carries ten times more bunch energy than that in an electron bunch potentially created in the proposed ILC.

Light electrons or positrons cannot be accelerated in circular accelerators due to substantial energy loss via synchrotron radiation, whereas the linear accelerators will be prohibitively long and costly to accelerate them to the TeV-scale energies. As a consequence, it has been suggested to transfer energies from proton bunches to the electron bunches via the beam-driven scheme. Owing to available competitive energy contents, proton bunches have the potential to power long distance wakefields and bring witness particles to the energy frontier in a single plasma stage. Caldwell et al. in 2009 [3, 4] first proposed the proton driven PWFA (PD-PWFA) scheme, and validated the acceleration of electrons to 0.62 TeV over a 450 m long uniform plasma channel. The average accelerating gradient is 1.38 GeV/m.

3.2.2 Dephasing Length

As a proton is 1837 times heavier than an electron or a positron, its relativistic factor is significantly lower given the same particle energy. Hence, in the proton driven plasma wakefield acceleration, there exists a velocity difference between the proton bunch and the accelerated electron or positron bunch. It results in a continuous and backward phase shift of the field pattern as a whole with respect to the witness bunch as the wake phase velocity is equal to the driving beam velocity. For a long distance beam-plasma interaction, the phase dephasing could be considerable thus should be taken into consideration.

The relative phase change ϕ between a proton driver and an accelerated electron bunch during a time T can be expressed as [5, 6]

$$\phi = k_p \int_0^T (v_e - v_p)dt, \tag{3.1}$$

where k_p is the wavenumber of the wakefield, v_e is the electron velocity and v_p is the proton velocity. To avoid the electron bunch slipping into the decelerating phase, ϕ must be smaller than π. The propagation distance corresponding to a phase slippage of π is called the dephasing length $L \simeq cT$.

The energy change of the proton and the electron reads

$$\frac{d(\gamma_p m_p c^2)}{dt} = -qE_{dec}v_p, \tag{3.2}$$

$$\frac{d(\gamma_e m_e c^2)}{dt} = eE_{acc}v_e, \tag{3.3}$$

where E_{dec} and E_{acc} are the decelerating amplitude seen by the proton and the accelerating amplitude seen by the electron, respectively, γ_p and γ_e are the relativistic factors of the proton and the electron, respectively. To simplify the calculation, we assume the wake structure does not evolve with time, i.e., the electrons and protons see constant wake amplitudes. Combing Eqs. 3.2 and 3.3, the phase change in an infinitesimal time interval Δt is

$$\Delta\phi = k_p(v_e - v_p)\Delta t = E_{wb}\left[\frac{\gamma_{e_{i+1}} - \gamma_{e_i}}{E_{acc}} + \frac{m_p e}{m_e q}\frac{\gamma_{p_{i+1}} - \gamma_{p_i}}{E_{dec}}\right], \tag{3.4}$$

where the subscripts i and $i+1$ denote the start and the end time of the time interval, $E_{wb} = m_e c\omega_p/e$ is the wave-breaking field.

The momentum equations for protons being decelerated and electrons being accelerated by the plasma wakefields are

$$\frac{d(\gamma_p m_p v_p)}{dt} = -qE_{dec}, \tag{3.5}$$

$$\frac{d(\gamma_e m_e v_e)}{dt} = eE_{acc}. \tag{3.6}$$

Similarly, in the time interval Δt, we can write

$$m_p c\left(\sqrt{\gamma_{p_{i+1}}^2 - 1} - \sqrt{\gamma_{p_i}^2 - 1}\right) = -qE_{dec}\Delta t, \tag{3.7}$$

$$m_e c\left(\sqrt{\gamma_{e_{i+1}}^2 - 1} - \sqrt{\gamma_{e_i}^2 - 1}\right) = eE_{acc}\Delta t. \tag{3.8}$$

As a result, $\gamma_{p_{i+1}}$ and $\gamma_{e_{i+1}}$ can be explicitly resolved given the earlier relativistic factors γ_{p_i} and γ_{e_i}. Then the phase shift during the time interval Δt is obtainable according to Eq. 3.4. The total phase dephasing is just addition of the phase shifts in all time intervals calculated in the iteration method.

After combing Eqs. 3.4, 3.7, and 3.8, it gives

$$\Delta\phi = \frac{E_{\mathrm{wb}}}{E_{\mathrm{acc}}}(\gamma_{e_{i+1}} - \gamma_{e_i})\left[1 - \frac{\left(\sqrt{\gamma_{e_{i+1}}^2 - 1} - \sqrt{\gamma_{e_i}^2 - 1}\right)(\gamma_{p_{i+1}} - \gamma_{p_i})}{\left(\sqrt{\gamma_{p_{i+1}}^2 - 1} - \sqrt{\gamma_{p_i}^2 - 1}\right)(\gamma_{e_{i+1}} - \gamma_{e_i})}\right]. \tag{3.9}$$

For relativistic electrons, $\gamma_{e_{i+1}} \gg \gamma_{e_i} \gg 1$, the above equation can be simplified to be

$$\Delta\phi = \frac{E_{\mathrm{wb}}}{E_{\mathrm{acc}}}(\gamma_{e_{i+1}} - \gamma_{e_i})\left[1 - \frac{(\gamma_{p_{i+1}} - \gamma_{p_i})}{\left(\sqrt{\gamma_{p_{i+1}}^2 - 1} - \sqrt{\gamma_{p_i}^2 - 1}\right)}\right]. \tag{3.10}$$

Now we can estimate the dephasing length given the initial beam and plasma parameters close to those used in the simulation introduced later. The initial proton energy is 1 TeV and the initial electron energy is 10 GeV. That is, $\gamma_{p_0} = 1063$ and $\gamma_{e_0} = 19515$. The plasma density is $5 \times 10^{14}\,\mathrm{cm}^{-3}$. E_{acc} and E_{dec} are taken as the average accelerating and decelerating gradients calculated based on the simulation results, which are $0.4E_{\mathrm{wb}}$ and $0.17E_{\mathrm{wb}}$, respectively. Figure 3.1 (the red line) shows that the dephasing length is 450 m where the electrons overrun the protons. This is obviously smaller than the simulated 700 m. The reasons are twofold. First, in the simulation, the protons located at different positions within the bunch see different decelerating gradients and get decelerated in different rates. While the low energy protons lag behind, giving rise to bunch elongation, the protons with high energies in the front still propagate in a larger velocity, leading a high wake velocity. In addition, the wakefield wavelength in the simulated case is much larger than the plasma wavelength as the hollow plasma channel is involved instead of the uniform plasma.

According to Eq. 3.1, the dephasing length mainly depend on v_{p}, that is, the proton energy and the decelerating gradient seen by protons, because the electron velocity v_{e} can be generally considered as the speed of light. When increasing the initial proton energy twice, the dephasing length increases significantly (Fig. 3.1 green line). Doubling the decelerating gradient brings the electrons to overrun the protons 65 m earlier (Fig. 3.1 blue line).

3.2.3 *Phase Mixing Effect*

From Chap. 2, we know for the positively and negatively charged beam drivers in the same condition, there is no amplitude difference between the plasma wakefields sourced in the linear regime but a phase shift of π. The witness bunch hence sees

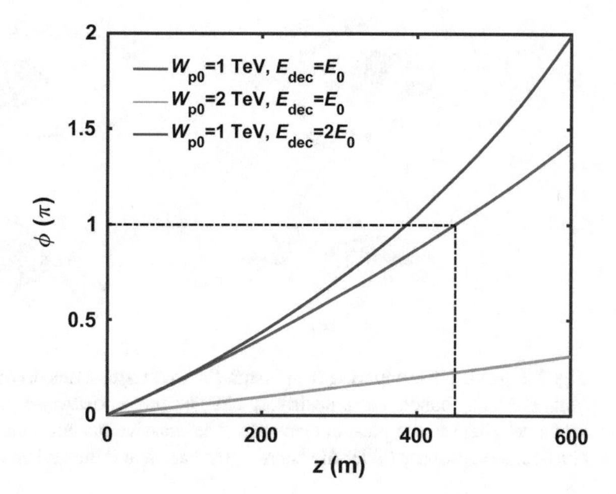

Fig. 3.1 Phase slippage versus the beam propagation distance. The red line denotes the case with the initial proton energy $W_{p0} = 1$ TeV and the decelerating gradient $E_{\rm dec} = E_0 = 0.17E_{\rm wb}$. The green line denotes the case with a twice larger initial proton energy. The blue line denotes the case with a twice larger decelerating gradient

exactly the same field. Nevertheless, the plasma responses to driving beams of different charge signs are asymmetric in the nonlinear regime. When a strong enough electron or laser beam traverses the plasma, the plasma electrons near axis are fully expelled away, leaving an electron-free bubble with purely uniform ions distributed. The bubble provides a linear focusing to the witness electron bunch, thus preserves its emittance. In this case, the plasma electrons initially feel the repulsive forces at the same time and thus oscillate at the same phase.

On the other hand, the positively charged beam (e.g., the proton bunch) attracts or sucks in plasma electrons from different radii, which then reach the axis at different times. The bubble shape is formed after the plasma electrons stream through the proton bunch. But since they leave the axis at different times, their oscillation phases are different. This is called the phase-mixing effect (see Fig. 3.2). It results in some plasma electrons trailing behind the proton bunch and hence a "non-clean" ion bubble. With nonuniform plasma electrons distributing in the bubble, the transverse wakefield is radially nonlinear and it also varies both in time and along the witness bunch. The nonlinear variations of radial forces could significantly disrupt the emittance of the accelerated beam and make it far from suitable for practical applications in future colliders. In addition, the less density compression on-axis due to the "suck-in" effect leads to a smaller accelerating gradient. This has been demonstrated in Ref. [7], where it further validates the promotion of the wake amplitude via a hollow plasma channel, which is a vacuum channel surrounded by the plasma annulus with a uniform density.

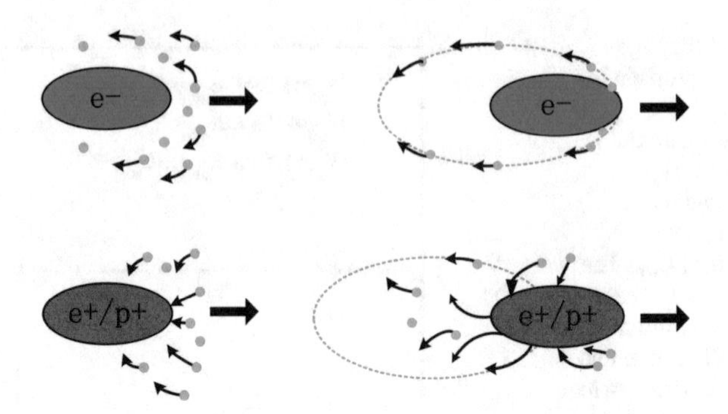

Fig. 3.2 A sketch comparing the plasma electron trajectories driven by a dense negatively charged (e.g., electron) bunch and a positively charged (e.g., positron or proton) bunch, respectively. The green dots denote the plasma electrons. The massive plasma ions are not shown here as they are considered stationary under the beam perturbation and thus taken as the uniform background

3.3 Hollow Plasma Channel

Above we have demonstrated that, despite the potential of accelerating particles to TeV-level energies in a single stage, proton-driven plasma accelerators are inferior in terms of conserving the beam quality compared to the electron or laser driven accelerators. Hollow plasma has been recognized to be capable of mitigating the beam quality deterioration, although it was initially proposed to confine the lasers [8]. With a hollow channel, an electron- and ion- free accelerating region with zero transverse forces is feasible, which benefits the preservation of the beam emittance [9].

Earlier theoretical studies of hollow plasma channels were mostly focused on low amplitude waves and linear plasma responses [7, 10–17] and on laser [10–14] or electron [15] drivers that push plasma electrons aside rather than pull them into the channel. Studies of positively charged drivers are fewer in number [7, 17–19], but a good positron acceleration regime was found for this configuration [19]. In this regime, the plasma response is strongly nonlinear, and a wide area devoid of both plasma electrons and ions exists in the channel, in which the accelerating field approaches the wave-breaking limit. In simulations, the witness positron beam not only obtained a high energy, but also preserved its normalized emittance and reached a final energy spread of 1.5%. The new regime opens a path to accelerating a considerable amount of positrons to TeV-range energies with low emittance and energy spread.

3.3.1 *Experimental Implementation*

In this section, we introduce the methods of implementing the hollow plasma channels, as it is indispensable to testify the concept of hollow plasma wakefield acceleration in experiment. Earlier work proposed to block the center of a UV laser beam

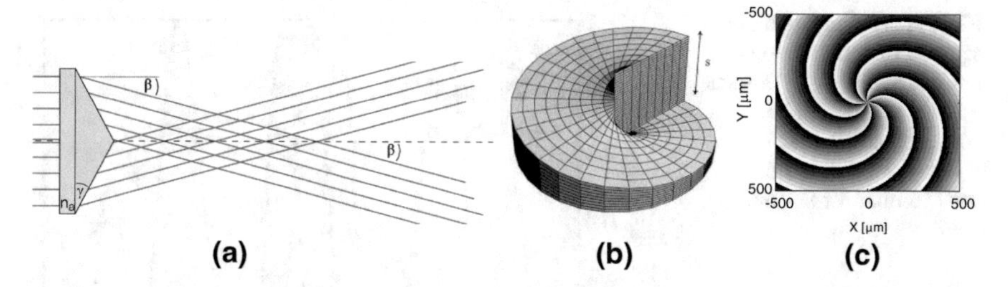

Fig. 3.3 **a** An axicon optic. The focused rays converge at the angle β. **b** A spiral phase optic. **c** A kinoform etching mask with 7 pieces of gratings ($m = 7$), each of which is formed by 8 etching layers in a staircase pattern. Images reproduced from Ref. [23]

with a mask so that only the vapor in the annular region was ionized if the laser intensity there was sufficient [20]. In this work, while the positron beam was successfully guided by the hollow channel, the channel was partially filled due to diffusion of the plasma as the positron beam arrived 200 ns later than the laser.

An advanced method is to combine an axicon lens (Fig. 3.3a) and a spiral phase lens (Fig. 3.3b) to form an aperture which shapes the phase and intensity of an incoming wavefront that is initially uniform in intensity and phase [21, 22]. The shaped laser then has a transverse intensity profile with the minimum intensity in the center but high intensity in a ring. As a result, the laser primarily ionizes the gas in the ring and a hollow plasma channel is formed.

An axicon, also referred as a Bessel optic, is a conically shaped lens which is capable of producing a long, tight focus called line-focus (the ray crossing region in Fig. 3.3a). It shifts the light by a phase that is linear along the radius (i.e., the distance to the optic center)

$$\Phi(r) = -k_\perp r, \tag{3.11}$$

where k is the laser wavenumber, $k_\perp = k \sin\beta$ is the perpendicular wavenumber, β is the convergence angle. Given a uniform illumination on the axicon with the incident intensity I_0, the intensity of the focused rays along the line-focus reads

$$I(r, z) = I_0 2\pi\beta^2 kz J_0^2(k_\perp r). \tag{3.12}$$

The transverse profile of the intensity follows the function $J_0^2(k_\perp r)$ (Fig. 3.4), indicating a maximum intensity on-axis.

To obtain desired minimum intensity on-axis, a spiral phase optic (see Fig. 3.3b) has been proposed to add angular momenta to the laser beam. The combined phase shift of an axicon and a phase screw is thus

$$\Phi_m(r, \phi) = -k_\perp r + m\phi, \tag{3.13}$$

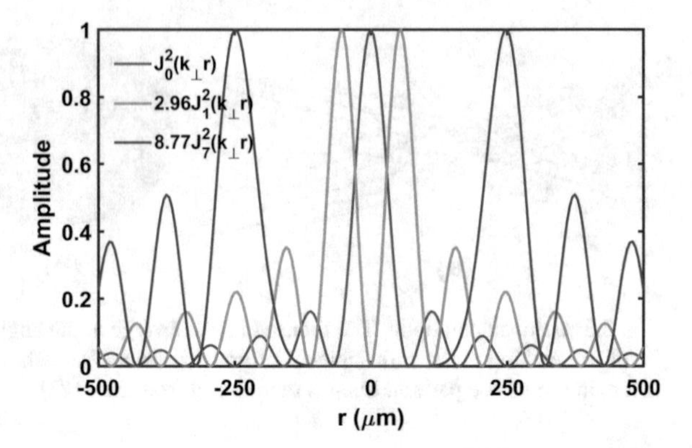

Fig. 3.4 A comparison of squared Bessel functions (or normalized intensity) versus r for $m = 0, 1, 7$. The convergence angle is $\beta = 4.4$ mrad. k is the wavenumber for 800 nm light

where the phase term $m\phi$ is introduced by the spiral optic. The resulting intensity along the line-focus of the axicon reads

$$I(r, z) = I_0 2\pi\beta^2 kz J_m^2(k_\perp r). \tag{3.14}$$

Hence, the transverse intensity profile goes as $J_m^2(k_\perp r)$. Figure 3.4 indicates that for $m > 0$ (high order Bessel), there exists an intensity minimum on-axis. The larger the m is, the wider the radial distance from the first maximum to the central minimum (i.e., the channel radius) is. Note that the amplitude of $J_m^2(k_\perp r)$ decreases with m, thus to obtain the same focused intensity, the incident intensity needs to be amplified by an coefficient (e.g., 2.96 for $m = 1$ and 8.77 for $m = 7$).

Equation 3.14 implies that the laser intensity at the focal region increases linearly along z. But this is not a big issue to produce a longitudinally uniform channel, thanks to the threshold nature of the ionization rate $W(s^{-1})$ which is estimated by the ADK model [24] and drops exponentially with the decreasing electric field. The ionization fraction is $F = W\tau$ when a laser pulse of length τ passes by. Given the laser pulse length, we can estimate the laser field threshold for an ionization fraction of 100%. Normally a slightly smaller field than the threshold will lead to orders of magnitude smaller ionization fraction [25], which indicates a sharp transition between the ionized region and the gas region in the radial direction. In the longitudinal direction, as long as the laser operates above the field threshold, the ionization is not sensitive to the small increase or variation of the laser intensity along the focus. Also further ionization of the ions is not a concern as it requires a significantly larger electric field. Hence, the hollow channel can be realized with a high longitudinal uniformity.

Although the combined optic (an axicon optic plus a spiral phase optic) performs an ideal intensity shape, its implementation in practice faces some issues. For instance, it is tricky to fabricate an accurate spiral screw due to the discontinuity at $r = 0$. In addition, the combined optic is very thick, leading to self-phase modulation of the high intensity laser pulses. Fortunately, a kinoform is developed which generates an approximate phase shift of the combined optic [22]. The kinoform optic is a phase plate etched with the spiral phase pattern following

$$\Phi(r, \phi) = \text{mod}(\Phi_m(r, \phi), \pi). \tag{3.15}$$

Figure 3.3c shows an etching mask for a kinoform optic at $m = 7$ where each grating is formed with 8 etching layers in a staircase pattern and each layer shifts the phase of the light by $\pi/4$.

By means of a kinoform, Fan et al. [26] obtained a laser beam with the transverse intensity following the fifth-order Bessel (J_5) profile and created a 0.8 cm long hollow plasma channel with the channel radius of $\sim$4 μm. A more encouraging plasma wakefield acceleration experiment [9] at FACET (Facility for Advanced aCcelerator Experimental Tests) of SLAC has demonstrated an 8 cm long hollow plasma channel with a radius of 250 μm, where the laser with a J_7 density profile ionized the lithium vapor $\sim$3 ps before the driving positron beam arrived, so the plasma diffusion can be neglected [27]. The channel radius here exactly agrees with the radial position of the first maximum calculated in Fig. 3.4. A follow-up experiment [28] created a 25 cm long hollow channel. This opens up prospects for practical applications of hollow channels in plasma wakefield acceleration.

3.4 Simulation Results

Future colliders need both high-energy and high-quality electrons and positrons. In this section, we explore the acceleration regime for electrons in a hollow plasma channel driven by a short proton bunch, which is complementary to the work of positron acceleration in Ref. [19]. It will be shown that the witness electron beam can simultaneously have a high charge and experience a strong accelerating field. This feature enables high energy transfer efficiencies, approaching those in the strong blowout regime in uniform plasmas driven by electrons [29]. Most importantly, we have discovered an equally good acceleration regime for electrons in hollow channels, despite that the similarity between electron and positron acceleration is not typical for nonlinear wakefields and even surprising. It contrasts with the uniform plasma case, for which the acceleration structure is strongly charge-dependent [30].

3.4.1 Concept of the Proposed Scheme and Simulation Parameters

In the uniform plasma, it is nearly impossible for the proton driver to create an electron-free blow-out area like the electron driver does, as the protons pull in the plasma electrons towards the propagation axis. The strongly nonlinear interaction

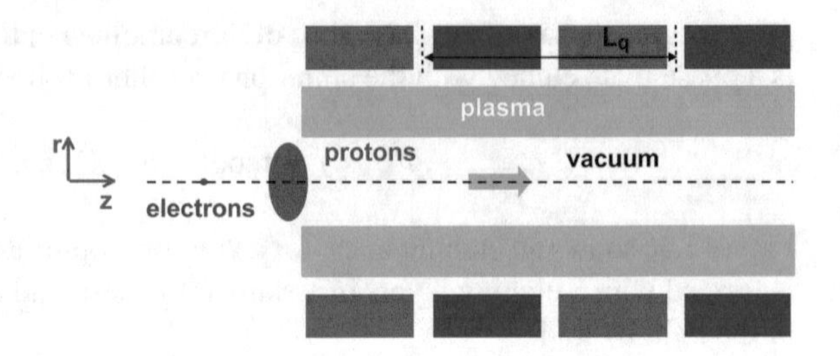

Fig. 3.5 Schematic of the concept of the proposed proton driven electron acceleration in a hollow plasma channel. The red and blue rectangles denote the quadrupole magnets with alternating polarity. L_q is the quadrupole period

might create a rarefied region of plasma electrons [3, 4, 31] but it requires the proton driver to be short enough to resonantly excite the plasma electron waves.

We propose a proton-driven hollow plasma scheme, which has the potential to realize an acceleration region devoid of plasma electrons. A simple schematic of this concept is illustrated in Fig. 3.5. The free plasma electrons and immobile ions are initially located outside a vacuum channel. When the proton bunch propagates through the hollow channel, it attracts plasma electrons into the channel and drives the wave, in which a considerable amount of electrons oscillates between the channel and the plasma. At certain wave phases, there are no electrons near the channel axis, but the longitudinal electric field is present. These regions are ideal for accelerating the witness beam [32], as the absence of both plasma electrons and ions results in no transverse wakefields and, thereby, conservation of the witness emittance. In the following, we demonstrate the efficient acceleration regime with simulations and discuss which of its specific features are responsible for particular advantages.

We use 2D axisymmetric quasi-static PIC code LCODE [33, 34] to conduct hundreds of meters long simulations owing to its high computational efficiency. In the simulations, both the beam and the plasma are modeled by fully relativistic macro-particles. The simulation window using the coordinate system (r, θ, ξ) travels at the speed of light c, where $\xi = z - ct$ is a co-moving coordinate. The window width is large enough (3 mm) to extend to the zero-field area, so the boundary conditions have no effect on the solution. The radial and axial grid sizes are $0.05c/\omega_p \approx 12\,\mu$m. The time step is $8.404\omega_p^{-1} \approx 6.7$ ps, corresponding to the length of 2 mm.

The parameter set of the simulated case is detailed in the Table 3.2. Most importantly, the driver radius ($\sigma_{rd} = 350\,\mu$m) is equal to the channel radius r_c, the driver length ($\sigma_{zd} = 150\,\mu$m) is slightly shorter than the skin depth of the outer plasma ($c/\omega_p = k_p^{-1} = 238\,\mu$m), the initial witness radius ($\sigma_{rw} = 10\,\mu$m) is much smaller than the channel radius, and the driver charge is approximately 20 times higher than the charge of electrons in volume k_p^{-3} of the unperturbed plasma. The witness charge amounts to 8.7% of the driver charge. Driver population (1.15×10^{11} protons) and energy ($W_{d0} = 1$ TeV) are intentionally chosen close to Refs. [3, 4, 19] to facilitate comparison.

The channel is surrounded by external quadrupole magnets, which define the axis of the system and keep the beams focused. The external quadrupoles were first proposed in Ref. [35] to precisely align the trajectory of the driver and to prevent the emittance driven erosion of the driver head. For short proton drivers that propagate

Table 3.2 Parameters for simulations

Parameters	Values	Units
Initial driving proton beam		
Proton population	1.15×10^{11}	
Initial energy, W_{d0}	1	TeV
Energy spread	10%	
RMS beam length, σ_{zd}	150	μm
RMS beam radius, σ_{rd}	350	μm
Angular spread	3.0×10^{-5}	
Initial witness electron beam		
Electron population	1.0×10^{10}	
Initial energy, W	10	GeV
Energy spread, $\delta W/W$	1%	
RMS beam length, σ_{zw}	15	μm
RMS beam radius, σ_{rw}	10	μm
Angular spread	1.0×10^{-5}	
Unperturbed hollow plasma		
Plasma density, n_p	5×10^{14}	cm^{-3}
Hollow radius, r_c	350	μm
Simulated plasma length	700	m
External quadrupole magnets		
Magnetic field gradient, S	500	T/m
Quadrupole period, L_q	0.9	m

hundreds of meters, the quadrupole focusing becomes a must [3]. In the hollow channel, not only the driver head but the whole driver and the witness are guided by quadrupoles, as there is no plasma focusing in the channel. According to Ref. [3], the quadrupole focusing works properly (i.e., the radial oscillations of the particles are insignificant) when the focusing period of quadrupoles is much shorter than the period of transverse particle oscillations, that is,

$$L_q \ll 2\pi\sqrt{\frac{W}{eS}}, \tag{3.16}$$

where L_q is the quadrupole period, W is the beam energy and S is the magnetic field gradient. For an achievable S of 0.5 T/mm and initial driver energy of W, the upper limit of L_q for the driver is 16 m. As the witness bunch has a lower initial energy, the corresponding upper limit is smaller than the one for the driver. Since the quadrupole magnets are installed along the whole plasma channel, their period should satisfy the smaller upper limit, i.e. the one for the witness bunch. In addition, the simulation indicates that L_q should be larger than 0.9 m in order to guarantee that over 98% of protons survive after 700 m propagation in the hollow plasma. We choose $S = 0.5$ T/mm and $L_q = 0.9$ m as the optimal quadrupole parameters which require a minimum quadrupole strength to sufficiently focus the driver. Based on the Eq. 3.16, the initial energy of the witness bunch must significantly exceed 3.1 GeV. 10 GeV is chosen to keep radial oscillations of the witness bunch small. Equation 3.16 also implies that the increase of the witness energy during the acceleration will gradually alleviate the limitation to the quadrupole period.

3.4.2 *Transverse Wakefield Characteristics and Beam Dynamics*

The spatial structure of the transverse plasma fields (Fig. 3.6a) in the efficient regime correlates with the density distribution of plasma electrons (Fig. 3.6c). The field structure in general resembles that of the bubble or blowout regimes in a uniform plasma driven by a laser or an electron beam, but here the drive and witness bunches reside in different bubbles. In the electron-free regions within the channel (blue area in Fig. 3.6c), the radial wakefield is exactly zero, as seen from the cyan area in Fig. 3.6a or the red line in Fig. 3.6b. Wherever plasma electrons enter the channel, the resultant force focuses protons and defocuses electrons (blue areas in Fig. 3.6a). This requires the witness bunch radius to be small.

The radial structure of the focusing force on the driver is similar to that of a sharp reflecting wall (the green line in Fig. 3.6b), which is beneficial in two aspects. First, the driver emittance does not blow up as the driver comes into the radial equilibrium [36, 37]. Second, as the energy W_d of driving particles decreases, the amplitude of their radial (betatron) oscillations remains constant and does not increase as $W_d^{-1/4}$, as it does in uniform plasmas [32, 38]. Consequently, driver depletion in energy does not result in driver widening and wakefield reduction. The strong defocusing outside the witness bunch region (see the red dashed line in Fig. 3.6a) makes external quadrupoles a necessary part of the proposed concept, as the quadrupoles control the witness bunch radius. Note that the external magnetic fields of the quadrupoles are not included in the transverse fields in Fig. 3.6. The quadrupoles also protect the driver head from emittance-driven erosion that would otherwise shorten the acceleration distance [4], as the plasma focusing is weak there (the blue line in Fig. 3.6b).

The witness bunch initially resides slightly behind the accelerating field maximum, so that the bunch head experiences a stronger field. This trick lowers the final energy spread, since the field slope at the witness location later changes its sign as the driver depletes. Due to strong nonlinearity of the plasma response, the transformer ratio is about unity.

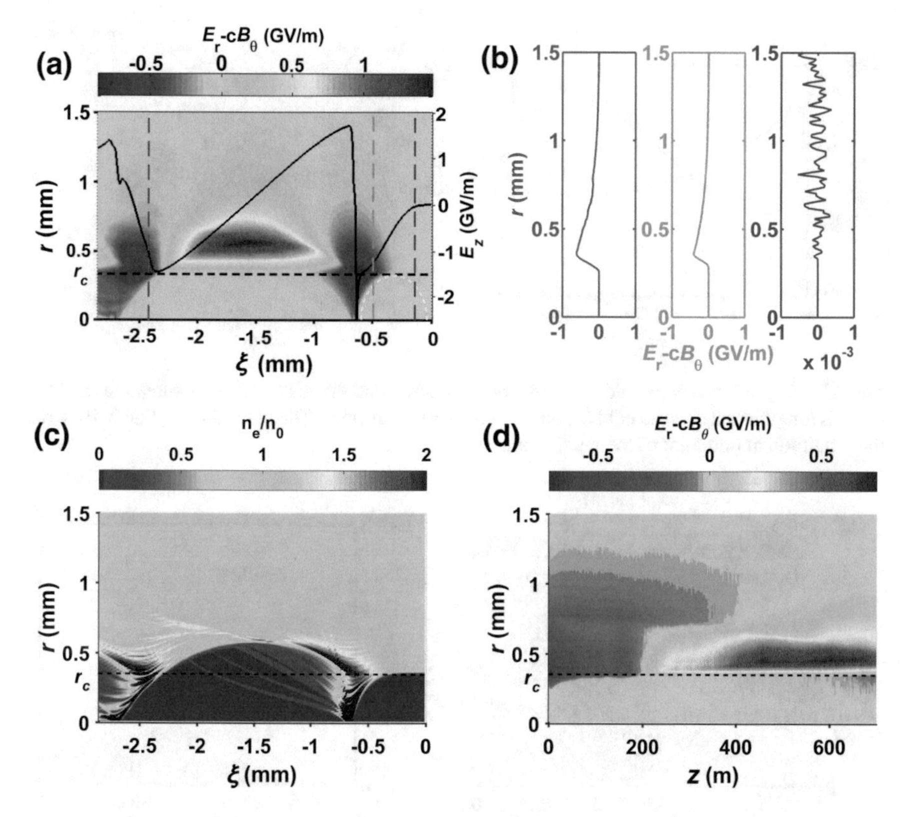

Fig. 3.6 **a** The transverse wakefield (colored map) and the on-axis longitudinal electric field (black line) at $z = 10$ m. The yellow dashed curve shows the driver location. The red dashed line is placed at the midpoint of the witness bunch. **b** The transverse wakefields at three longitudinal positions marked in **a** with vertical dashed lines of the corresponding colors. **c** The perturbed plasma density distribution at $z = 10$ m. **d** Evolution of the transverse wakefield at the midpoint of the witness bunch with the propagation distance. The horizontal black dashed lines mark the hollow channel radius. The simulation windows (a) and (c) travel to the right at the speed of light

3.4.3 Longitudinal Wakefield and Acceleration Characteristics

The final energy gain of the witness bunch (0.62 TeV $\approx 0.6\,W_{d0}$) and the required acceleration length (700 m) are comparable to the best results of Refs. [4, 19] (Fig. 3.7a). The beam energy increases linearly in the initial 400 m, afterwards the increase of the bunch energy slows down. But still, the average accelerating gradient exceeds 1.0 GeV/m over the first 600 m. The decrease of the acceleration rate mainly comes from the backward shift of the field pattern as a whole, which can be seen from Fig. 3.7b. This is caused by the relatively low relativistic factor of the driver in comparison with the witness electron bunch and is unavoidable for driver energies

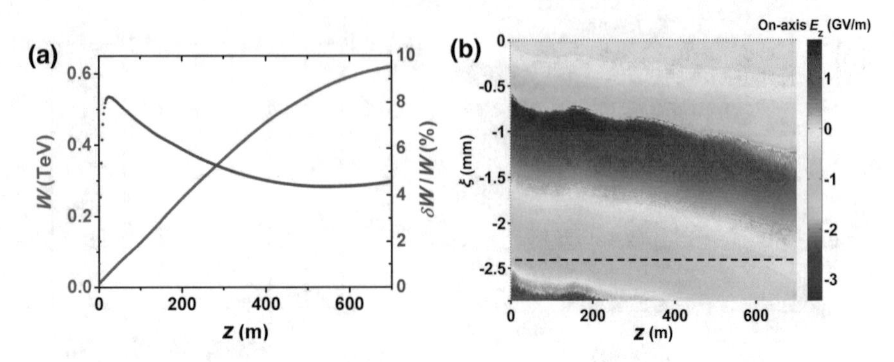

Fig. 3.7 Dependence of the mean witness bunch energy and the relative energy spread (a) and the on-axis longitudinal electric field (b) on the propagation distance. The black dashed line in **b** marks the longitudinal midpoint of the witness bunch

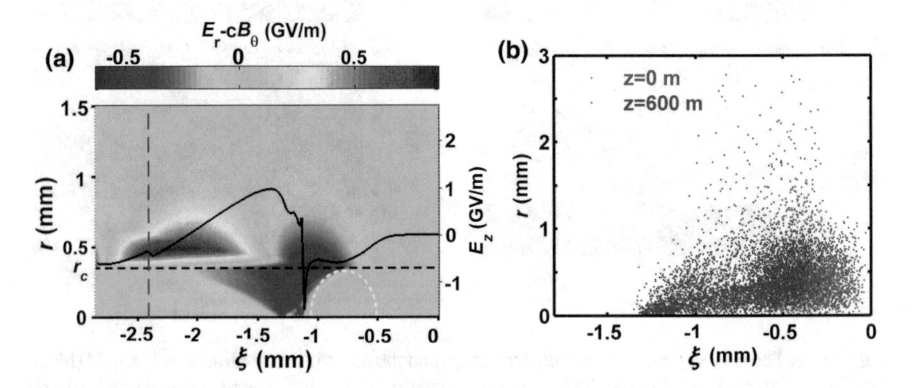

Fig. 3.8 **a** The corresponding transverse wakefield and the on-axis longitudinal electric field at $z = 600$ m in comparison with Fig. 3.6a. **b** Snapshots of the proton bunch at the beginning (red) and after a propagation distance of 600 m (blue)

of about 1 TeV or below [4]. However, since the witness bunch initially resides at the rear of the second bubble, the acceleration continues until the wave phase shifts back by almost a half of the period (Fig. 3.8a). At this point, the shape of the drive bunch has changed significantly (Fig. 3.8b). The bunch elongation results from the energy depletion and low energy protons lagged behind. Nevertheless, since driving protons cannot escape from the channel, the driver still excites relatively strong wakefields. The final energy spread of the witness bunch in this particular case is 4.6% (Fig. 3.7a). Note that here no optimization of the witness bunch shape has been done in order to reduce the energy spread.

The energy depletion of the proton bunch brings the plasma bubble shift backwards with respect to the witness bunch. Hence, the witness bunch sees more plasma electrons attracted into the channel appearing at large radii and forming a defocusing region (light blue area in Fig. 3.6d within the channel). Fortunately, with quadrupoles, the witness bunch is confined well within a small radial region, free from the dilution

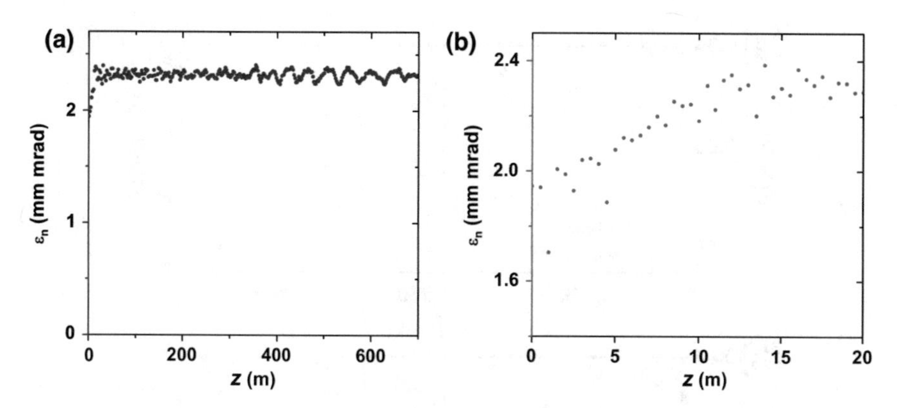

Fig. 3.9 **a** The normalized emittance of witness electrons over the whole acceleration process. **b** A close-up of the emittance growth in the first 20 m

of the plasma electrons. Therefore, the witness bunch feels no transverse plasma wakefield during the whole acceleration process.

As expected, the normalized emittance of the witness bunch is preserved at a low level of 2.4 mm mrad (Fig. 3.9a). To our knowledge, this is the first time the emittance preservation at this small level has been directly demonstrated in simulations of plasma wakefield acceleration over such long distances. This proves applicability of available PWFA simulation codes for studies of low-emittance beams. Apparently, if the initially loaded emittance is reduced, a smaller final emittance is obtainable owing to no blow-up arising from the transverse plasma wakefields.

Figure 3.9b indicates that the beam emittance increases from the initial 1.95 mm mrad to 2.4 mm mrad during the first 15 m. The emittance growth is caused by the quadrupoles, as there are no transverse plasma wakefields acting on the witness. The initial witness radius (10 μm) is larger than the equilibrium radius [4] (7.6 μm) calculated based on the quadrupole strength we use. During the first few meters of acceleration, the beam energy spread quickly blows up (Fig. 3.7a) and causes frequency variations in betatron oscillations and phase mixing. The incoherence of particle oscillations results in the emittance growth. But after 15 m, when the beam matches with the focusing structure, the normalized emittance levels off. We deliberately present the initially unmatched case here to demonstrate the scale of this effect.

3.4.4 Parameter Dependence and Reduction of the Energy Spread

The simulated case is typical for a wide range of parameter space of the hollow plasma. For example, Fig. 3.10 shows that variations of the channel radius from 320 μm to 380 μm or the plasma density from $4 \times 10^{14}\,\text{cm}^{-3}$ to $6 \times 10^{14}\,\text{cm}^{-3}$ do

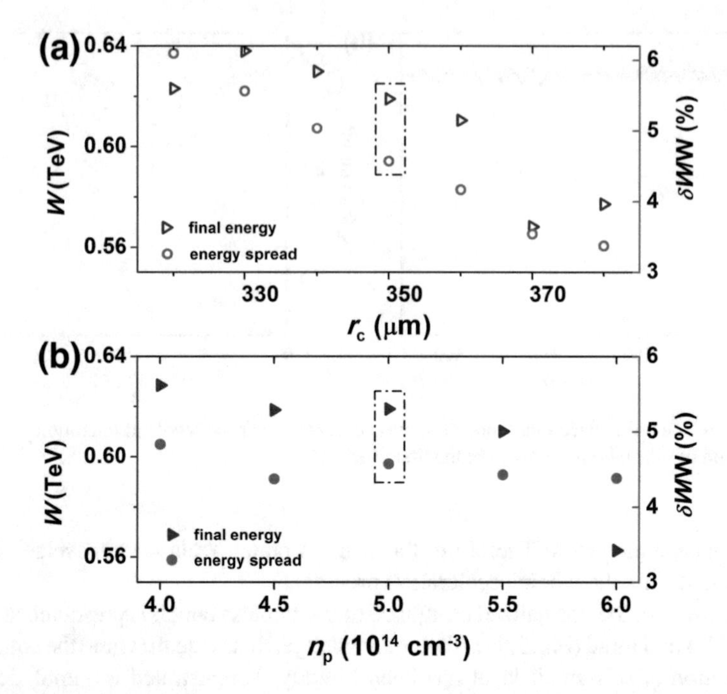

Fig. 3.10 Dependence of the energy gain and energy spread on the channel radius (a) and the plasma density (b). The dash-dotted frames denote the baseline case. The acceleration length is individually optimized for each data set

not result in substantial changes of the energy gain and energy spread. Furthermore, theoretically as long as the acceleration regime is free from plasma electrons and ions, the normalized emittance of the witness beam can be preserved at a low level, and the majority of electrons can survive with proper focusing from the external quadrupoles regardless of the plasma channel parameters. This is confirmed by Fig. 3.11.

In all aforementioned simulations, we have assumed that the hollow plasma channel has a sharp transition between the surrounding uniform plasma and the vacuum in the centre. Nevertheless, in practice, if the hollow channel is created by a shaped ionization laser, there is a moderate or even negligible fraction of gas ionization at the boundary where the laser electric field is very close to the ionization field threshold, as discussed in Sect. 3.3.1. As a result, there exist some finite density gradients at the boundary of the hollow channel. Figure 3.12a demonstrates different radial plasma density profiles where we simply assume linearly increasing plasma densities at the boundary in different rates. Apparently, despite various plasma density transitions at the boundary, the witness bunch is still accelerated to the same mean energy with an identical acceleration rate (Fig. 3.12b). The reason is straightforward. As the proton driver is intense, the slight density imperfection at the hollow channel boundary hardly alters the whole plasma density distribution significantly, thus the wakefield

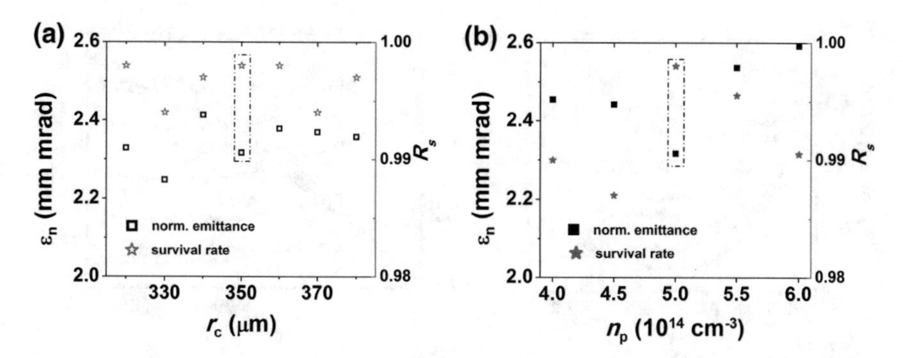

Fig. 3.11 Dependence of the normalized emittance and survival rate on the channel radius (a) and the plasma density (b). The dash-dotted frames denote the baseline case. The survival rate R_s here is defined as the ratio of the number of electrons present in the final accelerated bunch to the number of initially injected electrons. The acceleration length is individually optimized for each data set

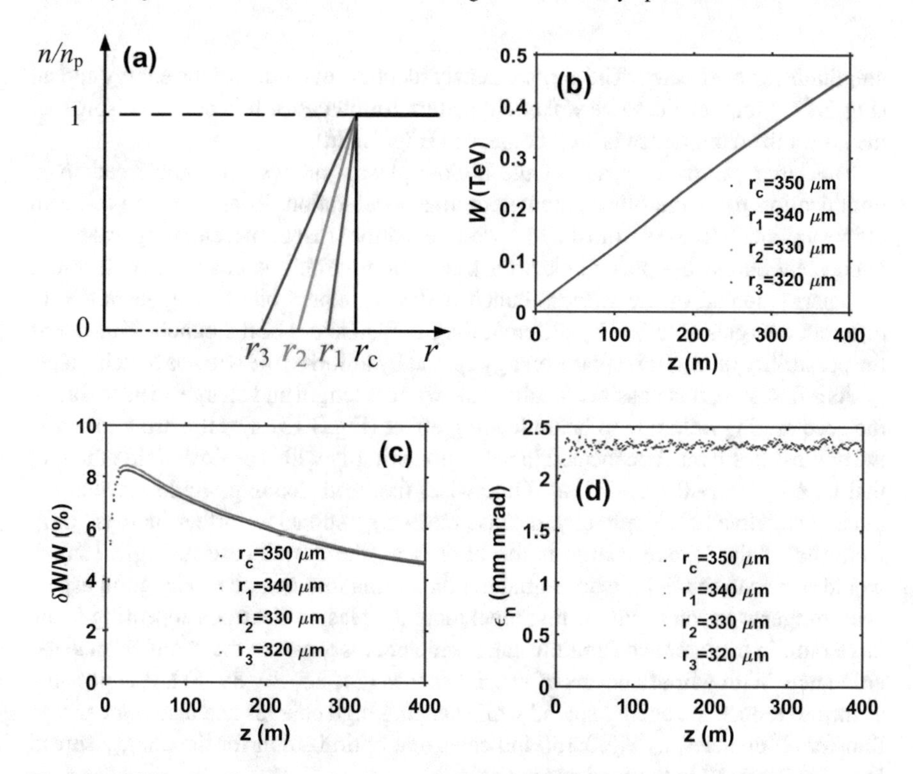

Fig. 3.12 Different radial plasma density profiles (a) and comparisons of the corresponding energy gain (b), normalized emittance (c) and energy spread (d) obtained by the witness electron bunch in different cases. The black colored case denotes the baseline case

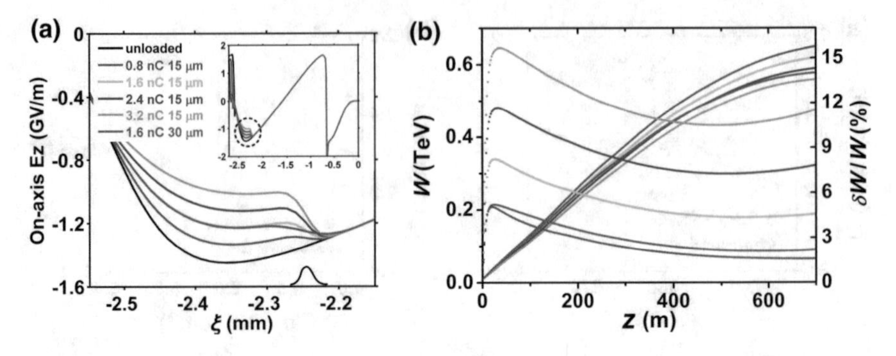

Fig. 3.13 **a** On-axis accelerating fields for unloaded and differently loaded cases with various charges and bunch lengths at $z = 10$ m. **b** Dependence of the mean energy and energy spread on the propagation distance for different witness bunches. The identically colored lines in **a** and **b** represent the same case. The proton driver and plasma parameters are the same as in Table 3.2

amplitude is not affected. The approximately identical evolution of the energy spread (Fig. 3.12c) implies the same wakefield pattern for all cases. It is not surprising that the normalized emittance is well conserved (Fig. 3.12d).

The large parameter space of the hollow plasma allows for some freedom in optimization for some other parameters, like acceleration distance, beam-to-beam efficiency, etc. It follows from the Maxwell's equations that the accelerating wakefield is constant across the plasma electron-free region within the channel. As a result, the energy spread of the witness bunch is the correlated one that appears due to different energies obtained by different longitudinal slices of the bunch. This opens the possibility of minimizing the energy spread by tailoring the witness bunch shape.

As a first step, it is possible to adjust the witness length and charge so as to flatten the accelerating field due to beam loading effect (Fig. 3.13). For this trick to work, the witness beam must be located in the region initially with a positive field gradient, that is, $\partial E_z/\partial z > 0$ (Fig. 3.13a). Otherwise, the beam loading would steepen the already negative field gradient and cause the energy spread to further increase [30]. Note that if the loaded charge is too high (e.g., the 3.2 nC case in Fig. 3.13a), it would overload the field, whose gradient then turns into negative. The good aspect is a stronger electron beam loading repel more the plasma electrons appearing in the large radii in the later acceleration stage, which helps preserve the beam normalized emittance. With witness beams of special shapes [14, 15, 29, 39–41], it is possible to further reduce the energy spread while keeping high charges and also high energy transfer efficiencies, as Fig. 3.13b indicates that optimization for the energy spread has a weak effect on the final energy gain.

3.4.5 *Comments on the Validity of Simulations*

In the simulations, we assume the beams propagate perfectly along the axis of the hollow channel. In general, hosing-like instabilities [42–45] might arise and destroy the beams when a transverse displacement of the beam couples with the displacement of the surrounding plasma. For our proposed acceleration scheme, the conducted two-dimensional axisymmetric simulations can not validate this effect, as full three-dimensional simulations are required for that. However, the experience gathered over decades of studying various beam-plasma instabilities suggests that this configuration is stable, because hosing-like instabilities are more a danger for long beams. If the beam is about one plasma wavelength long or even shorter, the hosing instability is suppressed [46, 47], and experimental observations of the stable propagation of short beams confirm this [2, 48]. Furthermore, large differences in oscillation frequencies suppress the coupling of transverse oscillations between the driver and the witness bunch [49], similar to the BNS damping (proposed by Balakin, Novokhatsky and Smirnov (BNS)) in conventional accelerating structures [50]. Because of this, even trains of many bunches can stably propagate in plasma wakefield accelerators [51]. Resonances between transverse particle oscillations and sign-varying pushes from the quadrupoles are taken into account in our simulation model [35] and do not result in loss of the witness beam quality. In Chap. 7, we will assess the beam deflection by the transverse fields induced by beam-channel misalignment through the 2D cartesian simulations, but note that the beam hosing does not appear therein.

For short beams (comparable to plasma wavelength or shorter), there is no difference between the 2D axisymmetric simulations and 3D simulations. Also note that the 2D simulations conducted here have reached the front line of simulation capabilities. The 3D simulations are hardly possible for modern codes.

3.5 Summary and Outlook

In this chapter, we propose a good regime for plasma wakefield acceleration of electrons in hollow channels driven by protons and illustrate it with simulations. The proposed scheme solves the issue of beam normalized emittance degradation due to strong, radially- and time-varying focusing forces in a uniform plasma driven by a positively charged beam. In this regime, the witness bunch resides in a region with a strong accelerating field and an absence of plasma electrons. The region preserves its shape and location up to driver depletion, thus providing a high acceleration efficiency and an average transformer ratio of about unity. The witness bunch takes advantage of the hollow channel; it experiences a radially uniform accelerating field and no transverse plasma fields. Therefore, the normalized beam emittance is conserved during acceleration, and the energy gain depends only on the longitudinal position along the bunch, thus enabling minimization of the energy spread by tailoring the bunch shape. The quadrupoles used to confine the driver from emittance-driven widening also play an important role in the confinement of the witness bunch. They keep the witness bunch from entering the defocusing region that appears at large radii as the driver depletes. The resultant high quality electron beam with our proposed scheme may find applications in the next generation energy frontier colliders.

The biggest challenge of assessing the proposed concept in experiment is the production of a long hollow channel, which is far from maturity given the fact that, so far the longest hollow channel implemented for plasma wakefield acceleration is only 25 cm long. However, irrespective of the variation of the channel thickness (width of the ionized plasmas) which is determined by the peak-to-peak spacing of the Bessel maxima, it is feasible to create a hollow channel with a length of at least 3 m [23]. In addition, the channel could be sectionalized into multiple plasma cells with almost no decrease of the longitudinal fields, provided that the gap between cells is shorter than the betatron period of driving particles [52].

References

1. Leemans W, Gonsalves A, Mao H-S, Nakamura K, Benedetti C, Schroeder C, Toth C, Daniels J, Mittelberger D, Bulanov S et al (2014) Multi-GeV electron beams from capillary-discharge-guided subpetawatt laser pulses in the self-trapping regime. Phys Rev Lett 113(24):245002
2. Blumenfeld I, Clayton CE, Decker F-J, Hogan MJ, Huang C, Ischebeck R, Iverson R, Joshi C, Katsouleas T, Kirby N et al (2007) Energy doubling of 42 GeV electrons in a metre-scale plasma wakefield accelerator. Nature 445(7129):741
3. Caldwell A, Lotov K, Pukhov A, Simon F (2009) Proton-driven plasma wakefield acceleration. Nat Phys 5(5):363
4. Lotov K (2010) Simulation of proton driven plasma wakefield acceleration. Phys Rev Spec Top-Accel Beams 13(4):041301
5. Ruth RD, Morton P, Wilson PB, Chao A (1984) A plasma wakefield accelerator. Part Accel 17(SLAC-PUB-3374):171
6. Xia G, Mete O, Aimidula A, Welsch C, Chattopadhyay S, Mandry S, Wing M (2014) Collider design issues based on proton-driven plasma wake-field acceleration. Nucl Instrum Methods Phys Res Sect A 740:173–179
7. Lee S, Katsouleas T, Hemker R, Dodd E, Mori W (2001) Plasma wakefield acceleration of a positron beam. Phys Rev E 64(4):045501
8. Katsouleas T, Chiou T, Decker C, Mori W, Wurtele J, Shvets G, Su J (1992) Laser wakefield acceleration & optical guiding in a hollow plasma channel. In: Proceedings of AIP conference, vol 279. AIP, pp 480–489
9. Gessner S, Adli E, Allen JM, An W, Clarke CI, Clayton CE, Corde S, Delahaye J, Frederico J, Green SZ et al (2016) Demonstration of a positron beam-driven hollow channel plasma wakefield accelerator. Nat Commun 7:11785
10. Schroeder C, Benedetti C, Esarey E, Leemans W (2016) Laser-plasma based linear collider using hollow plasma channels. Nucl Instrum Methods Phys Res Sect A 829:113–116
11. Chiou T, Katsouleas T, Decker C, Mori W, Wurtele J, Shvets G, Su J (1995) Laser wake-field acceleration and optical guiding in a hollow plasma channel. Phys Plasmas 2(1):310–318
12. Shvets G, Wurtele J, Chiou T, Katsouleas TC (1996) Excitation of accelerating wakefields in inhomogeneous plasmas. IEEE Trans Plasma Sci 24(2):351–362
13. Schroeder CB, Esarey E, Benedetti C, Leemans W (2013) Control of focusing forces and emittances in plasma-based accelerators using nearhollow plasma channels. Phys Plasmas 20(8):080701

14. Schroeder C, Benedetti C, Esarey E, Leemans W (2013) Beam loading in a laser-plasma accelerator using a near-hollow plasma channel. Phys Plasmas 20(12):123115
15. Chiou T, Katsouleas T (1998) High beam quality and efficiency in plasma based accelerators. Phys Rev Lett 81(16):3411
16. Schroeder C, Whittum D, Wurtele J (1999) Multimode analysis of the hollow plasma channel wakefield accelerator. Phys Rev Lett 82(6):1177
17. Kimura W, Milchberg H, Muggli P, Li X, Mori W (2011) Hollow plasma channel for positron plasma wakefield acceleration. Phys Rev Spec Top-Accel Beams 14(4):041301
18. Yi L, Shen B, Lotov K, Ji L, Zhang X, Wang W, Zhao X, Yu Y, Xu J, Wang X et al (2013) Scheme for proton-driven plasma-wakefield acceleration of positively charged particles in a hollow plasma channel. Phys Rev Spec Top-Accel Beams 16(7):071301
19. Yi L, Shen B, Ji L, Lotov K, Sosedkin A, Wang W, Xu J, Shi Y, Zhang L, Xu Z et al (2014) Positron acceleration in a hollow plasma channel up to TeV regime. Sci Rep 4:4171
20. Marsh K, Blue B, Clayton C, Joshi C, Mori W, Decker F, Hogan M, Iverson R, O'Connell C, Raimondi P et al (2003) Positron beam propagation in a meter long plasma channel. In: Proceedings of particle accelerator conference, vol 1. IEEE, pp 731–733
21. Andreev N, Aristov YA, Polonskii LY, Pyatnitskii L (1991) Bessel beams of electromagnetic waves: self-effect and nonlinear structures. Sov Phys JETP 73(6):969
22. Andreev NE, Bychkov SS, Kotlyar VV, Margolin LY, Pyatnitskii LN, Serafimovich P (1996) Formation of high-power hollow Bessel light beams. Quantum Electron 26(2):126–130
23. Gessner SJ (2016) Demonstration of the a hollow channel plasma wakefield accelerator. PhD thesis, Stanford University, Palo Alto, California, USA
24. Ammosov M, Delone N, Krainov V, Perelomov A, Popov V, Terent'ev M, Yudin GL, Ivanov MY (1986) Tunnel ionization of complex atoms and of atomic ions in an alternating electric field. Sov Phys JETP 64(6):1191–1194
25. Bruhwiler DL, Dimitrov D, Cary JR, Esarey E, Leemans W, Giacone RE (2003) Particle-in-cell simulations of tunneling ionization effects in plasma-based accelerators. Phys Plasmas 10(5):2022–2030
26. Fan J, Parra E, Alexeev I, Kim K, Milchberg H, Margolin LY, Pyatnitskii L (2000) Tubular plasma generation with a high-power hollow Bessel beam. Phys Rev E 62(6):R7603
27. Chen FF (2006) Introduction to plasma physics and controlled fusion, volume 1: plasma physics. Springer, Berlin
28. Lindstrfim C, Adli E, Allen J, An W, Beekman C, Clarke C, Clayton C, Corde S, Doche A, Frederico J et al (2018) Measurement of transverse wake-fields induced by a misaligned positron bunch in a hollow channel plasma accelerator. Phys Rev Lett 120(12):124802
29. Lotov K (2005) Efficient operating mode of the plasma wakefield accelerator. Phys Plasmas 12(5):053105
30. Lotov K (2007) Acceleration of positrons by electron beam-driven wakefields in a plasma. Phys Plasmas 14(2):023101
31. Corde S, Adli E, Allen J, An W, Clarke C, Clayton C, Delahaye J, Frederico J, Gessner S, Green S et al (2015) Multi-gigaelectronvolt acceleration of positrons in a self-loaded plasma wakefield. Nature 524(7566):442
32. Assmann R, Yokoya K (1998) Transverse beam dynamics in plasma-based linacs. Nucl Instrum Methods Phys Res Sect A 410(3):544–548
33. Lotov K (2003) Fine wakefield structure in the blowout regime of plasma wake-field accelerators. Phys Rev Spec Top-Accel Beams 6(6):061301
34. Sosedkin A, Lotov K (2016) LCODE: a parallel quasistatic code for computationally heavy problems of plasma wakefield acceleration. Nucl Instrum Methods Phys Res Sect A 829:350–352
35. Kudryavtsev A, Lotov K, Skrinsky A (1998) Plasma wake-field acceleration of high energies: physics and perspectives. Nucl Instrum Methods Phys Res Sect A 410(3):388–395
36. Lotov K (1998) Simulation of ultrarelativistic beam dynamics in plasma wake-field accelerator. Phys Plasmas 5(3):785–791

37. Lotov K (2017) Radial equilibrium of relativistic particle bunches in plasma wakefield accelerators. Phys Plasmas 24(2):023119
38. Lotov K (1998) Simulation of ultrarelativistic beam dynamics in the plasma wake-field accelerator. Nucl Instrum Methods Phys Res A 3(410):461–468
39. Katsouleas TC, Wilks S, Chen P, Dawson JM, Su JJ (1987) Beam loading in plasma accelerators. Part Accel 22:81–99
40. Tzoufras M, Lu W, Tsung F, Huang C, Mori W, Katsouleas T, Vieira J, Fonseca R, Silva L (2008) Beam loading in the nonlinear regime of plasma based acceleration. Phys Rev Lett 101(14):145002
41. Tzoufras M, Lu W, Tsung F, Huang C, Mori W, Katsouleas T, Vieira J, Fonseca R, Silva L (2009) Beam loading by electrons in nonlinear plasma wakes. Phys Plasmas 16(5):056705
42. Whittum DH, Sharp WM, Simon SY, Lampe M, Joyce G (1991) Electron-hose instability in the ion-focused regime. Phys Rev Lett 67(8):991
43. Whittum DH, Lampe M, Joyce G, Slinker SP, Simon SY, Sharp WM (1992) Flute instability of an ion-focused slab electron beam in a broad plasma. Phys Rev A 46(10):6684
44. Lampe M, Joyce G, Slinker SP, Whittum DH (1993) Electron-hose instability of a relativistic electron beam in an ion-focusing channel. Phys Fluids B: Plasma Phys 5(6):1888–1901
45. Krall J, Joyce G (1995) Transverse equilibrium and stability of the primary beam in the plasma wake-field accelerator. Phys Plasmas 2(4):1326–1331
46. Dodd E, Hemker R, Huang C-K, Wang S, Ren C, Mori W, Lee S, Katsouleas T (2002) Hosing and sloshing of short-pulse GeV-class wakefield drivers. Phys Rev Lett 88(12):125001
47. Huang C, Lu W, Zhou M, Clayton C, Joshi C, Mori W, Muggli P, Deng S, Oz, T Katsouleas E et al (2007) Hosing instability in the blow-out regime for plasma-wakefield acceleration. Phys Rev Lett 99(25):255001
48. Hogan M, Barnes C, Clayton C, Decker F, Deng S, Emma P, Huang C, Iverson R, Johnson D, Joshi C et al (2005) Multi-GeV energy gain in a plasma-wakefield accelerator. Phys Rev Lett 95(5):054802
49. Mehrling T, Fonseca R, de la Ossa AM, Vieira J (2017) Mitigation of the hose instability in plasma-wakefield accelerators. Phys Rev Lett 118(17):174801
50. Balakin V, Novokhatsky A, Smirnov V (1983) VLEPP: transverse beam dynamics. In: Proceedings of the 12th international conference on high energy accelerators, vol 830811, pp 119–120
51. Lotov K (1998) Instability of long driving beams in plasma wakefield accelerators. In: Proceedings of the 6th European particle accelerator conference, Stockholm, 1998, pp 806–808
52. Adli E (2016) Towards AWAKE applications: electron beam acceleration in a proton driven plasma wake. In: Proceedings of IPAC16, Busan, Korea, JACOW, Geneva, Switzerland, pp 2557–2560

Chapter 4
Multiple Proton Bunch Driven Hollow Plasma Wakefield Acceleration in Nonlinear Regime

4.1 Introduction

Proton-driven plasma wakefield acceleration has been demonstrated in simulations to be capable of accelerating particles to the energy frontier in a single stage, but its potential is hindered by the fact that currently available proton bunches are orders of magnitude longer than the plasma wavelength. Fortunately, proton micro-bunching allows driving plasma waves resonantly (discussed in Sect. 4.2.2). In this chapter, we propose using a hollow plasma channel for multiple proton bunch driven plasma wakefield acceleration and demonstrate that it enables the operation in the nonlinear regime and resonant excitation of strong plasma waves. This new regime also involves beneficial features of hollow channels for the accelerated beam (such as emittance preservation and uniform accelerating field) and long buckets of stable deceleration for the drive beam.

In the following we first explain why we study multiple proton bunches driving a hollow plasma. Then we introduce the simulated case used for illustrating the multi-proton bunch driven regime. We elucidate the characteristics of nonlinear wakefields driven in the hollow plasma and the behaviour of multiple proton bunches in the wakefield potential wells. Afterwards we demonstrate the acceleration characteristics and conservation of beam quality of the witness bunch. Finally, we discuss the dependence of results on four important quantities: plasma skin depth, driver radius, hollow channel radius and micro-bunch period, and assess the effect of longitudinal plasma density inhomogeneity on the acceleration.

Y. Li, *Studies of Proton Driven Plasma Wakefield Acceleration*, Springer Theses,
https://doi.org/10.1007/978-3-030-50116-7_4

4.2 Evolution of the Proposed Concept

4.2.1 Difficulties of Currently Producible Proton Bunches in Use

Previous studies on proton driven plasma wakefield acceleration have assumed the proton bunch as short as hundreds of μm long [1–5], whereas presently producible high-energy proton bunches are tens of cm long. For example, the SPS operates the proton bunch at an RMS length of 12 cm. The LHC runs 7 TeV proton bunches with the RMS bunch length of 7.6 cm. The reason is in synchrotrons the proton beam usually runs for tens of to hundreds of million turns to get huge energies. Hence, it is a must to keep the beam long so that the transverse and longitudinal beam instabilities are relaxed and the beam can stay stably in the synchrotrons for long times [6–8]. The longitudinal momentum noise is intentionally introduced to the beam during acceleration. In doing so, the beam lengthens during acceleration and its longitudinal space expands. The beam peak current decreases and stays below the thresholds of the instabilities.

The problem is that, such proton bunches produced in synchrotrons are much longer than the usual submillimeter plasma wavelength, therefore they hardly excite strong plasma wakefields directly. It is conceivable to longitudinally compress the bunch length to the plasma wavelength by traditional methods [9–12], among which is the typical magnetic chicane. It has been popular in use for bunch compression in free electron lasers. It consists of a linear RF section to introduce an energy chirp to the beam and a set of dipoles to create a dispersive path for the particle path modulation. To be specific, the beam is initially injected off-crest in the RF accelerating field, so that the beam tail particles obtain more energies than the beam head particles. When passing the dipoles, the high energy particles are deflected less, hence they take shorter paths and catch up with the low energies particles in the beam front. In this way, the bunch is compressed when exiting from the magnetic chicane. However, as the proton bunch has ultra-high energy and the required compression factor is up to several orders of magnitude, the bunch compression is too costly and technically challenging to implement. For example, Ref. [12] numerically calculates that it requires a roughly 600 m long compression system to shrink the 450 GeV, 12 cm SPS bunch to 3.6 mm, of which the RF system operating at 720 MHz is 500 m long to introduce a sensible energy spread given that the field gradient sustained in the RF cavity is 25 MeV/m. In addition, the system is roughly estimated up to a few km long if it is to compress an LHC bunch, which is prohibitive.

4.2.2 Generation of a Proton Bunch Train

Fortunately, the self-modulation instability (SMI) [10, 13–15] offers a new solution to cope with this issue. The SMI is the seeded axisymmetric mode of the transverse

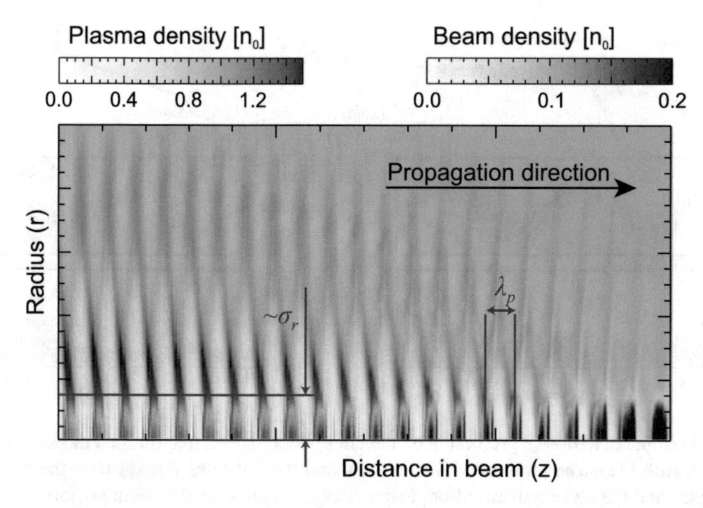

Fig. 4.1 Example of a train of proton micro-bunches generated via self-modulation instability, reproduced from Ref. [21]. The z coordinate here denotes the beam position, different from the one we use below to represent the propagation distance. λ_p is the plasma wavelength. σ_r is the beam RMS radius

two-stream instability [16, 17] that develops in the beam-plasma system. The transverse wakefield of the beam arising from the noise is amplified by a proper seeding. It periodically focuses and defocuses different slices of the beam, leading to a rippling of the beam, which further amplifies the plasma waves. Because of this positive feedback, the long proton bunch is transformed into a train of micro-bunches that follow equidistantly at the plasma wavelength [18–20], as shown in Fig. 4.1. With this multiple micro-proton bunch driven scheme, it has demonstrated the capability of accelerating an electron bunch to the multi-TeV scale with the existing LHC bunch [13]. The downside is protons located between the microbunches are deflected by the plasma defocusing forces and form a halo out of the bunch train. The proton halo hardly contributes to the wake excitation, which accounts for a majority of proton energy loss. But still, in comparison with the traditional bunch compression, this plasma-based modulation is essentially cheap and easy, hence ideally suitable for the initial studies.

A more efficient approach to create a sequence of short proton bunches is the longitudinal density modulation, the mechanism of which resembles that of the longitudinal bunch compression mentioned above, except that the energy chirping operates at a much higher frequency. Reference [22] demonstrates the basic idea, where a mm-wavelength dielectric accelerating structure is used to bring energy modulation to the long proton beam, then the energy modulation is transformed into proton micro-bunching with the beamline of magnetic dipoles as the proton path in the dipoles linearly depends on the proton momentum (Fig. 4.2). It has been calculated that, with a 5 m long dielectric accelerator whose operation frequency is 300 GHz and field gradient is over 100 MeV/m, plus a 120 m long magnetic beamline, the SPS

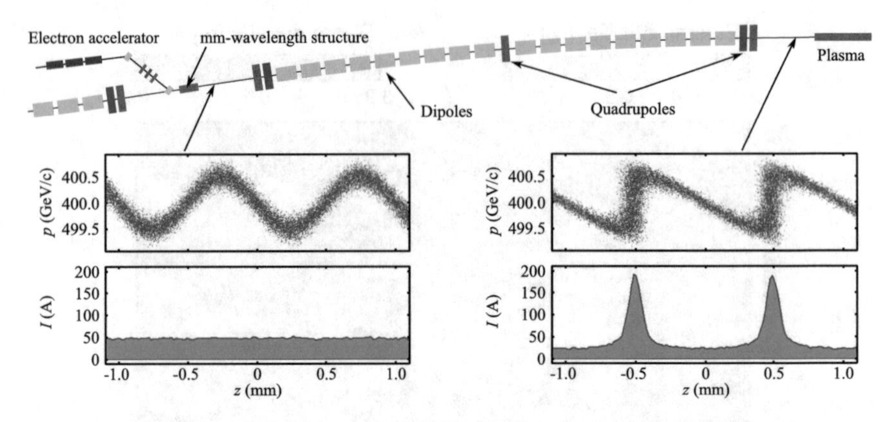

Fig. 4.2 Example of a train of proton micro-bunches generated via longitudinal modulation, reproduced from Ref. [22]. The plots in the second row and the third row demonstrate the beam longitudinal space and the current distribution, respectively. z represents the beam position

bunch is transformed into a train of 1 mm long short proton bunches. Apparently, the longitudinal modulation is substantially less challenging than the bunch compression in creating efficient proton bunches for the plasma wakefield acceleration. Also more protons remain in use in comparison with the transverse modulation based on SMI.

4.2.3 Necessity of a Hollow Plasma Channel

While the SMI forces the micro-bunching of the long proton bunch and allows harnessing high energy protons in the plasma wakefield acceleration, it intrinsically limits the beam-plasma interaction to the linear or weakly nonlinear regimes [13, 23, 24]. This is because the wave period elongates as nonlinear effects come into play, and the wave eventually falls out of resonance with the bunch train [23]. For the pre-formed proton bunch train via the longitudinal modulation, assuming a first proton bunch drives nonlinear wakefields, then in the trailing bubble areas the ions are more prevalent than the plasma electrons, which determines them defocusing for the subsequent proton bunches if injected. Therefore, the multiple proton bunches can only stably operate in the linear regime in uniform plasma. However, in the linear regime, the plasma focusing on the witness bunch strongly depends on its density. For a high witness bunch charge, the focusing nonlinearly varies both across and along the witness bunch [25]. The accelerating field also inevitably varies radially instead of being constant as in the nonlinear regime, which is detrimental to the witness beam quality [26].

The hollow plasma channel [27] is generally one of two effective solutions to the deterioration issues of witness beam quality discussed above. Its features have been demonstrated in Chap. 3. Another solution is the strongly nonlinear (blowout) regime in which the witness bunch propagates in a "bubble" void of plasma electrons, but containing plasma ions [28, 29]. The blowout regime is easier to implement experimentally, thus it has been extensively studied during the last two decades [30, 31]. However, it has several potential drawbacks. First, very dense particle beams required for collider applications can produce large perturbations in the ion density, giving rise to nonlinear transverse fields inside the bubble and degradation of the witness beam quality [32]. Although massive ions with heavier elements are less prone to density perturbations, they incur a prohibitively large growth of witness emittance from multiple Coulomb scattering [33, 34]. Most importantly, the bubble naturally appears for laser or electron drivers, but not for protons or positrons. There is a blowout-like regime for positively charged drivers [1, 2, 35], but the witness bunch quality degrades more due to the presence of plasma electrons in the bubble.

The hollow plasma channel is free from the aforementioned drawbacks, which stimulates excessive theoretical studies even though experimental implementation of such a channel is still in its infancy [36]. Up to now, all theoretical studies of hollow channels were dedicated to short drivers (both for particle beams [3–5, 37–40] and for lasers [27, 41–47]). In particular, an advantageous nonlinear regime has been discovered for short proton bunches of a high charge [3–5], whereas these bunches are difficult to produce. In this chapter, we investigate whether equally good acceleration conditions are possible with trains of short, lower charge proton bunches.

4.3 Simulations of the Proposed Regime

4.3.1 Considerations of Beam and Plasma Parameters

We propose to introduce ten identical and equidistant proton bunches to an axisymmetric hollow channel of radius r_c. The surrounding plasma outside the channel is of uniform density n_p. The bunches initially have the density distribution

$$n_b = \frac{N_b}{2(2\pi)^{3/2}\sigma_r^2\sigma_z} e^{\frac{-r^2}{2\sigma_r^2}} \left[1 + \cos\left(\sqrt{\frac{\pi}{2}}\frac{\xi - \xi_i}{\sigma_z}\right)\right], \; |\xi - \xi_i| < \sqrt{2\pi}\sigma_z \tag{4.1}$$

where σ_r and σ_z are the RMS bunch radius and length, N_b is the bunch population, and ξ_i is the centroid of the ith bunch, where $\xi = z - ct$. The shifted cosine shape is a convenient approximation to the Gaussian shape with the RMS length of σ_z, but without infinitely tails. The bunch train period is $\approx 10\sigma_z$. The electron witness bunch has a similar shifted cosine shape and is initially positioned beyond the maximum accelerating field to extend the acceleration distance [5]. The witness charge and shape are not optimized for the minimum energy spread, as we do not intend to demonstrate the capabilities of the optimal beam loading. The initial witness radius is much smaller than the channel radius, which is necessary for the sake of emittance

Table 4.1 Parameters for simulations

Parameters	Values	Units
Initial proton beam		
Population of a single bunch, N_b	1.15×10^{10}	
Energy, W_{d0}	1	TeV
Energy spread	10%	
Single bunch length, σ_z	63	μm
Beam radius, σ_r	71	μm
Angular spread	5×10^{-5}	
Bunch train period	631	μm
Initial witness electron beam		
Population	2×10^9	
Energy, W	10	GeV
Energy spread, $\delta W/W$	1%	
Bunch length, σ_{zw}	15	μm
Bunch radius, σ_{rw}	10	μm
Normalized emittance, ε_n	2	mm mrad
Unperturbed hollow plasma		
Plasma density, n_p	6×10^{15}	cm^{-3}
Hollow radius, r_c	200	μm
External quadrupole magnets		
Magnetic field gradient, S	500	T/m
Quadrupole period, L_q	0.9	m

conservation. Still, it is not as small as required for collider applications, since this would require reducing the simulation grid size and increasing the simulation time impractically.

In order to resonantly drive strong plasma waves, the bunches are supposed to be in tune with the wakefields and reside in the decelerating phases. Presence of the hollow channel results in longer wakefield wavelength than in uniform plasma. Given the bunch period, the wider the channel, the larger the required plasma density. The channel must be wide enough to accommodate most of the driver particles, so we choose $r_c \approx 3\sigma_r$ and adjust the plasma density to fit the resonance between the bunches and wakefields. Similarly to Refs. [1–5], we surround the plasma by focusing quadrupole magnets, which keep the driver head from emittance-driven erosion and control the witness electron trajectories.

The beam and plasma parameters are given in detail in Table 4.1. A typical energy of 1 TeV is chosen as the driver energy. Such high energy also allows us to verify the long term (150 m in our simulated case) bunch propagation and excitation of nonlinear wakes. The simulations are performed with the 2D LCODE [48, 49] using the cylindrical geometry (r, θ, ξ).

4.3.2 Nonlinear Wakefield Excitation in Hollow Plasma

In the proposed regime, most driver protons are initially located within the hollow channel radius of 200 μm (Fig. 4.3a). Protons that start inside the surrounding plasma experience a positive radial force (Fig. 4.3c) and are quickly defocused (Fig. 4.3e). This results in some change of the plasma wave amplitude driven in the first few meters of propagation (Fig. 4.3a). In addition, the wave amplitude increases with the distance from the driver head. This leads to nonlinear elongation of the wakefield period [23]. Hence the wake is not strictly periodic. Nevertheless, since the elongation is insignificant, all bunches still reside in the decelerating phases and see rising decelerating gradients on average.

In view of the perturbation of plasma electron density (Fig. 4.3b) and the bunch portraits (Fig. 4.3e), we see that the first driving bunch stays in a weakly nonlinear wake and the subsequent bunches stay in almost spherical plasma bubbles and experience nonlinear waves of increasing amplitude. The maximum radius of each bubble is larger than the hollow channel radius. It suggests the existence of areas out of the hollow channel but within the plasma electron sheath, where the uniformly distributed plasma ions are more prevalent than the plasma electrons being pulled there. These areas correspond to the yellow and red regions in Fig. 4.3c denoting positive transverse wakefields. These positive fields will defocus the protons if they get there. The blue area within the channel in Fig. 4.3c imposes focusing on the protons. It is created by plasma electrons attracted into the channel, whose trajectories are seen in Fig. 4.3b.

The proton bunches reside in rear halves of the bubbles and are almost half a period long. The bunches completely fill decelerating phases of the wave (Fig. 4.3a) and nevertheless stably propagate along the channel over a long distance (Fig. 4.3e), which indicates that the bunches are strongly focused. This contrasts to wave characteristics in uniform plasmas, where regions of simultaneous deceleration and focusing are as short as a quarter of the wave period for low-amplitude waves [50] and even shorter for nonlinear waves and positively charged beams [30, 51]. Unique focusing ability of the nonlinear proton-driven wave is attributed to the basin-like radial profile of the wakefield potential in the channel (Figs. 4.3d and 4.4), i.e., the potential drops with the decreasing radius within the channel (Fig. 4.4d). In turn, this particular potential structure appears as protons experience a negative (focusing) or zero radial force all through the channel (Fig. 4.3c). Regions of strong defocusing force, which are necessary for potential oscillations and excitation of longitudinal fields, are located outside the channel and thus have no effect on the proton bunches.

Let

$$W_t = p_r^2/(2\gamma_b) + \Delta\Psi(r, \xi), \tag{4.2}$$

where $p_r^2/(2\gamma_b)$ denotes the transverse kinetic energy of a proton (p_r is its radial momentum and γ_b is its relativistic factor), and $\Delta\Psi(r, \xi)$ represents the potential energy difference between the value at the proton position and the maximum value

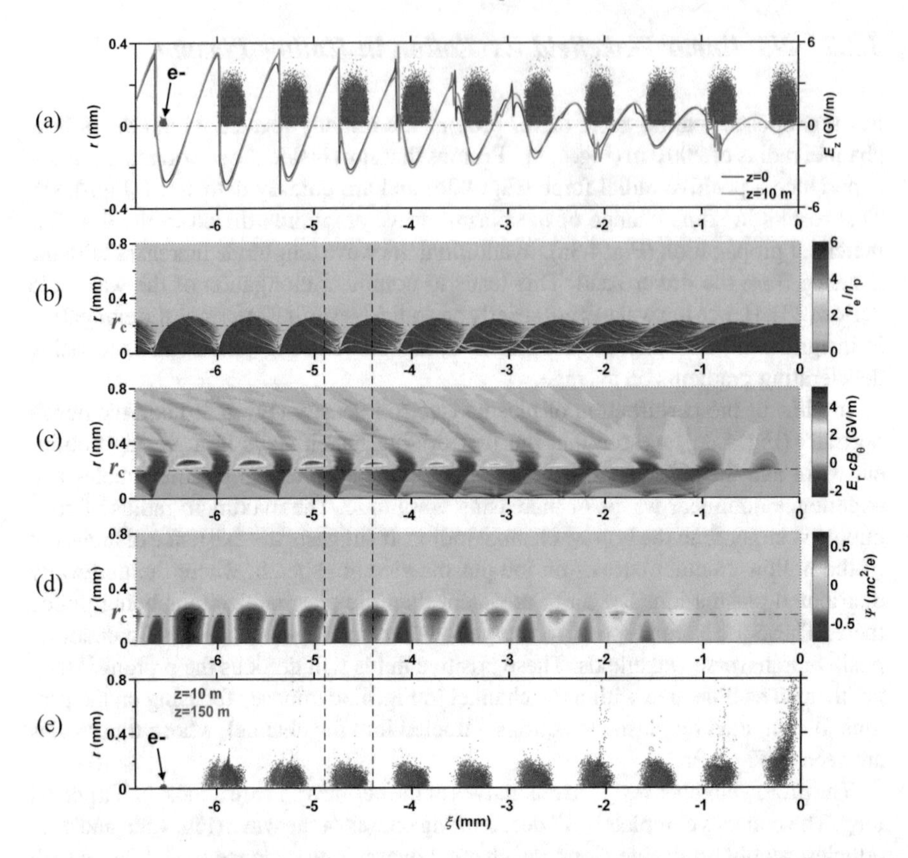

Fig. 4.3 **a** Initial distribution of 10 proton driving bunches and the witness electron bunch in the real space and the on-axis longitudinal electric fields at $z = 0$ and $z = 10\,\text{m}$, respectively. The plasma electron density (b), transverse plasma wakefields (c) and wakefield potential (d) at $z = 10\,\text{m}$. **e** Snapshots of the driver and the witness bunches at $z = 10\,\text{m}$ and $z = 150\,\text{m}$, respectively. The red dash-dotted lines in **b**, **c** and **d** illustrate the hollow channel boundary. Two vertical black lines are sketched to facilitate the discussion in the following. The observation window travels to the right at the speed of light

along the radius (at the basin edge). If $W_t > 0$, the proton has enough kinetic energy to escape from the potential well. Otherwise it stays in the well, that is, inside the channel. Most of driver protons have $W_t < 0$ all the interaction time (red dots in Fig. 4.4). More than 90% of protons in total survive over 150 m (Fig. 4.5). Small proton loss occurs at the boundary areas between positive and negative potentials, where the magnitude of $\Delta\Psi(r, \xi)$ is very small due to the lack of plasma electrons inside the channel (like the green line in Fig. 4.4d). This loss can also be observed in Fig. 4.3e where a narrow slice of protons escapes out of the hollow channel. Another reason for the observed proton loss is unphysical and related to the final width of the simulation window. There is no plasma focusing at the head of the first bunch, so the protons perform large radial oscillations under the weak focusing of external

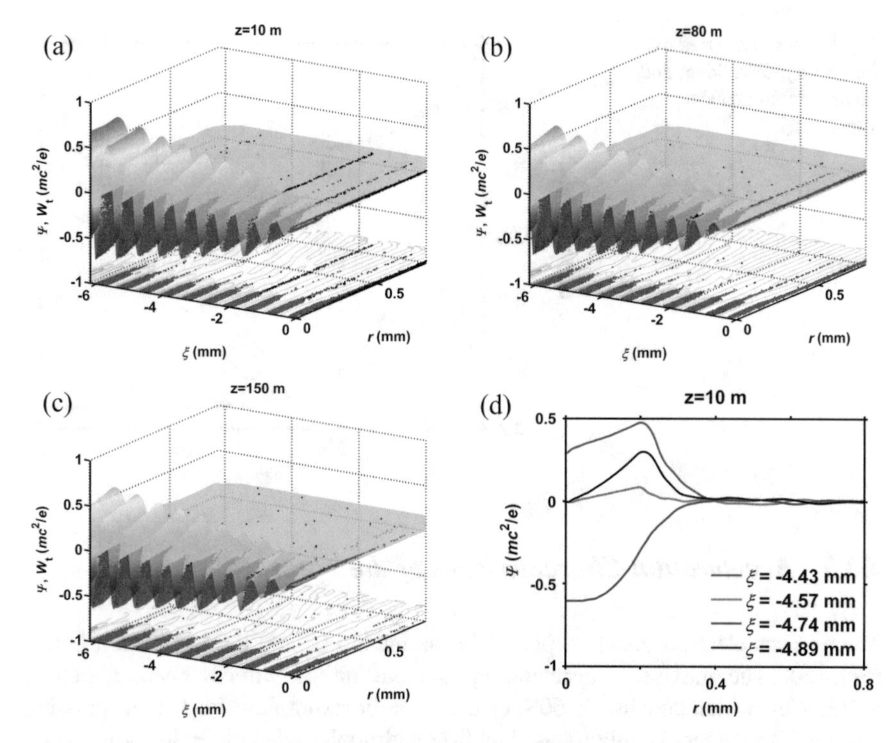

Fig. 4.4 Spatial distributions (a, b, c) of wakefield potential Ψ (surface, $\Psi(r,\xi)=\int_{-\infty}^{\xi} E_z(r,\xi')\mathrm{d}\xi'$) and driver protons (red dots for trapped and black for untrapped ones) at different propagation distances. The spatial positions of the protons are determined by the coordinates (r, ξ, W_t). **d** Radial dependencies of the potential at different ξ-positions ($z=10\,\mathrm{m}$) where the on-axis potential has local extrema or zero, i.e., $\xi=-4.43\,\mathrm{mm}$ (maximum), $\xi=-4.57\,\mathrm{mm}$ (zero), $\xi=-4.74\,\mathrm{mm}$ (minimum), and $\xi=-4.89\,\mathrm{mm}$ (zero). The two vertical black lines in Fig. 4.3 mark the range $\xi\in[-4.89-4.43]$ discussed here

quadrupoles (Fig. 4.6a). Some of the protons exit the simulation window radially and are counted as lost in Fig. 4.5. In our case, protons in the first bunch being excluded this way account for 2.8% of the total driver charge.

Protons at the back of the first bunch experience stronger plasma focusing, thus oscillating with a higher betatron frequency and within the hollow channel (Fig. 4.6b) in comparison with the head protons. Figure 4.6c gives the trajectory of a proton in the seventh bunch initially located in a fairly deep potential well. As a result, the proton oscillates well within 200 μm with a even higher betatron frequency. It is worth pointing out that the oscillation period changes insignificantly with the propagation distance in each proton case as the proton energy (or the relativistic factor) drops moderately.

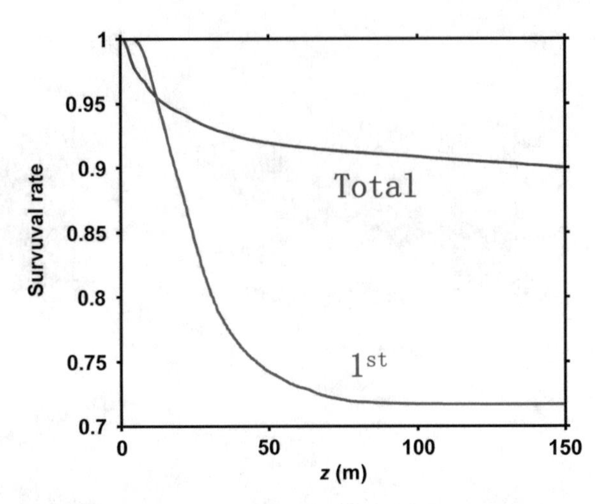

Fig. 4.5 Survival rates for the whole proton driver and the first driving bunch, respectively

4.3.3 Acceleration Characteristics of the Witness Bunch

The witness electron bunch is placed in the bubble behind the last proton bunch (Fig. 4.3). The maximum accelerating gradient in the witness beam region is 4.4 GeV/m, which constitutes 60% of the wave-breaking field for the surrounding plasma. The witness is only focused by the quadrupoles which keep its radius small and prevent it from being defocused by the plasma electrons staying in larger radii. Therefore, in our studied scheme, long intervals of proper focusing for the positively charged driver and for the electron witness beam coexist. The electron bunch is initially placed behind the peak accelerating field (see Fig. 4.3a) so that it can stay in the accelerating phase for a long distance. But the accelerating field initially has a negative slope, which leads to a large increase of the energy spread. When the bunch slips into the accelerating region with a positive field slope due to the backward shift of the wake phase, the energy spread decreases. After being accelerated over 150 m, the witness bunch reaches an energy of 0.47 TeV with an energy spread as low as 1.3% (Fig. 4.7a). The normalized emittance of the witness bunch stays at the initial level of 2 mm mrad over this distance (Fig. 4.7b). After that the witness beam quality starts to degrade because the depleted driver bunches change their shapes, which causes mismatch of the wave and the bunch train, penetration of plasma electrons into the witness beam region, and defocusing of witness electrons.

The trajectory of a witness electron is shown in Fig. 4.6d. The electron oscillates within 10 μm, and the oscillation period increases as the energy grows. Since witness electrons only experience weak quadrupole focusing and the amplitude and frequency of betatron oscillations are small, the betatron radiation can be foreseen to be modest.

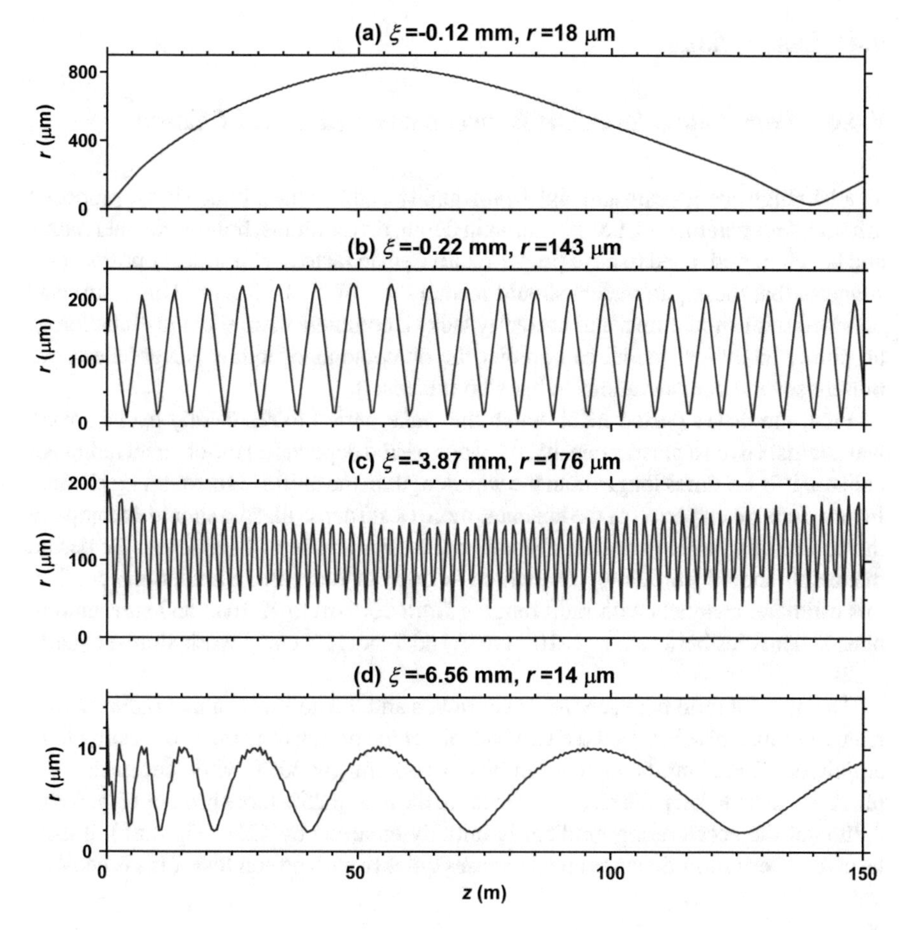

Fig. 4.6 Trajectories of protons initially located at the head (a) and back (b) of the first driving bunch and at the front (c) of the seventh driving bunch and the trajectory of the witness electron (d)

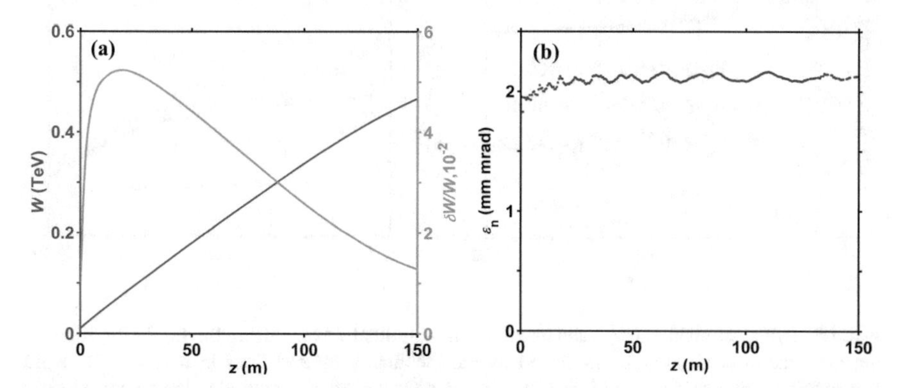

Fig. 4.7 **a** Energy (red line) and correlated energy spread (green line); **b** normalized emittance (blue points) of the witness bunch with respect to the acceleration distance

4.4 Discussions

4.4.1 Parametric Relation Between the Beam and Plasma

To attain high energy gain and high beam quality of the witness bunch in the proposed scheme, four parameters, i.e., plasma skin depth, driver radius, hollow channel radius and bunch period, need to have proper relative scale factors. Numerical optimization suggests that these parameters should form a ratio of $1 : 1 : 3 : 3\pi$. This result was obtained with simulations and currently lacks a complete theoretical justification as no purely analytical model can predict the outcome as of today. Nevertheless, we present several considerations to back up this result.

First, the driver period must match the wake period to resonantly excite strong wakefields. Given a plasma density, the wake period depends on the channel radius. In our case it is 1.5 times longer than the wavelength in the uniform surrounding plasma. In turn, for a given period of the bunches, more (less) dense plasma should accompany the choice of a larger (smaller) channel radius. For a train of N bunches, the period mismatch should be below 1/N, which is 10% in our case. Simulations confirm this estimate: channels with radii ranging from 160 μm to 220 μm and surrounding plasma densities between $5 \times 10^{15}\,\text{cm}^{-3}$ and $7 \times 10^{15}\,\text{cm}^{-3}$ work almost equally well.

The optimal ratio between the driver radius and the hollow channel radius comes from a compromise between the survival rate of the protons and the accelerating field amplitude. Reducing the channel radius to 160 μm (by 20%) while decreasing the plasma density to keep the resonance leads to the loss of 25% more protons (Fig. 4.8a). Although the accelerating gradient is initially enlarged by 42% (Fig. 4.8b), it only lasts for a very short distance and decreases quickly with proton loss. This is because

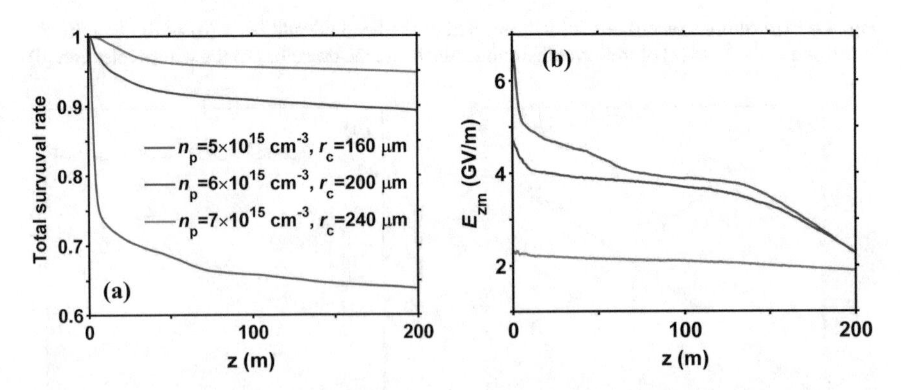

Fig. 4.8 Driver survival rates (a) and maximum longitudinal wakefield amplitudes (b) for different plasma structures. Each case is denoted by the identically colored lines in **a** and **b**. To avoid contributions from field singularities (Fig. 4.3a) into the wakefield amplitudes, the on-axis electric field E_z is first integrated along ξ and then multiplied by the wake wavenumber and averaged in wide (1 m) intervals in z

protons in large radii out of the hollow boundary easily escape due to positive radial wakefields. When increasing the channel radius to 240 μm, we obtain a survival rate as high as 95% (Fig. 4.8a), but the gradient decreases considerably (Fig. 4.8b).

The optimal ratio between the plasma skin depth and hollow channel radius aims at providing a strong wakefield on the axis, presence of plasma electrons in most cross-sections of the channel, and availability of plasma electron-free bubbles for hosting the witness bunch. This ratio depends on the charge of driver bunches and needs to be adjusted for each particular driver. If the plasma density is lower than the optimal one and the skin-depth (or the plasma wavelength) is longer (Fig. 4.9a), the driver charge increases as the bunch length extends. There are fewer plasma electrons in the channel, which is unfavorable for driver focusing. Hence, a large quantity of protons get lost, which subsequently destroys the accelerating field (Figs. 4.9b, c). On the other hand, if the plasma density is higher, the bubble structure disappears, bunch contributions stop adding up, and the accelerating field diminishes (Figs. 4.9d, e, f).

4.4.2 Effect of Plasma Density Inhomogeneity on Acceleration

To get a deeper insight into the effect of channel density inhomogeneity, we introduce a perturbation to the density of the surrounding plasma as

$$n = n_0(1 + \delta n \sin(2\pi z/L), \tag{4.3}$$

where $n_0 = 6 \times 10^{15}\,\text{cm}^{-3}$ is the baseline plasma density, δn is the perturbation amplitude and L is the perturbation period. The wakefield characteristics in different cases ($L = 0.1\,\text{m}$, $1\,\text{m}$ and $\delta n = 0.5\%$, 2.5%, 5%) are illustrated in Fig. 4.10. In addition, an irregular density non-uniformity following the form

$$n = n_0\left(1 + 0.025\left(1 - \cos\left(\frac{2\pi z}{1\,\text{m}}\right)\right) + 0.015\sin\left(\frac{2\pi z}{3\,\text{m}}\right) + 0.015\sin\left(\frac{2\pi z}{21\,\text{m}}\right)\right) \tag{4.4}$$

is introduced to further validate the proposed scheme.

The conclusions drawn from Fig. 4.10 are fourfold. First of all, the proposed scheme can tolerate a regular density perturbation up to a level of 5% and is slightly more sensitive to the irregular perturbation (Fig. 4.10b). Second, the oscillations of the maximum longitudinal wakefields follow the perturbation periods (Fig. 4.10a) and reducing the periods below the proton oscillation period can significantly boost the average wakefield amplitudes (the green line in Fig. 4.10b). Furthermore, a rise of the plasma density generally damps the wakefields while a drop of the plasma density increases the wakefields (Fig. 4.10a) similarly to the uniform plasma case [52]. Last but not least, a "clean" accelerating region for the witness beam remain under the plasma density perturbation discussed here and thereby its normalized emittance is

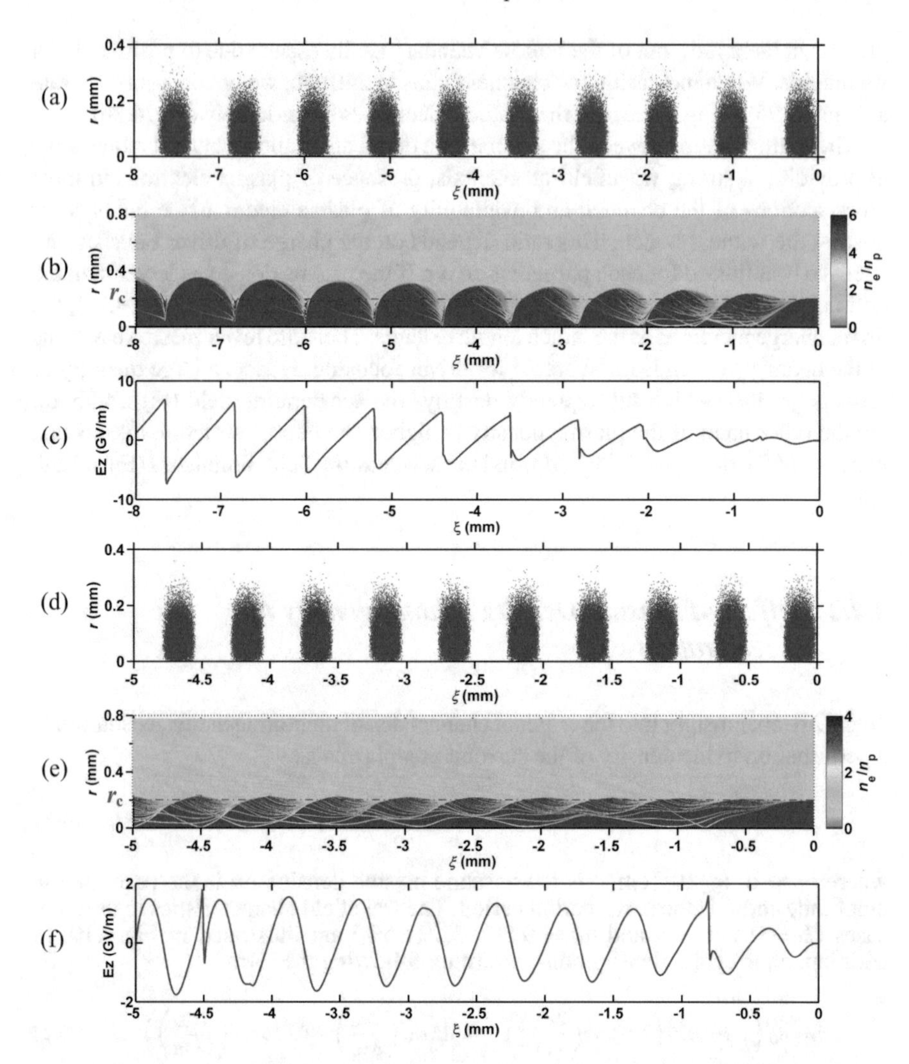

Fig. 4.9 Initial driver distribution in the real space (a, d), plasma electron density (b, e), and on-axis longitudinal electric field (c, f) for plasma densities of 3×10^{15} cm^{-3} (a, b, c) and 12×10^{15} cm^{-3} (d, e, f). Driver radius, hollow channel radius, and peak driver density are the same as in the baseline variant; the driver period is adjusted to match the wake period, so is the bunch length. The red dash-dotted lines in **b** and **e** denote the hollow channel boundary

well preserved (Fig. 4.10c). As a consequence, the limitation on tolerable density perturbations comes from decreasing the acceleration rate (Fig. 4.10d), rather than degrading the witness beam quality as in the case of uniform plasmas [52].

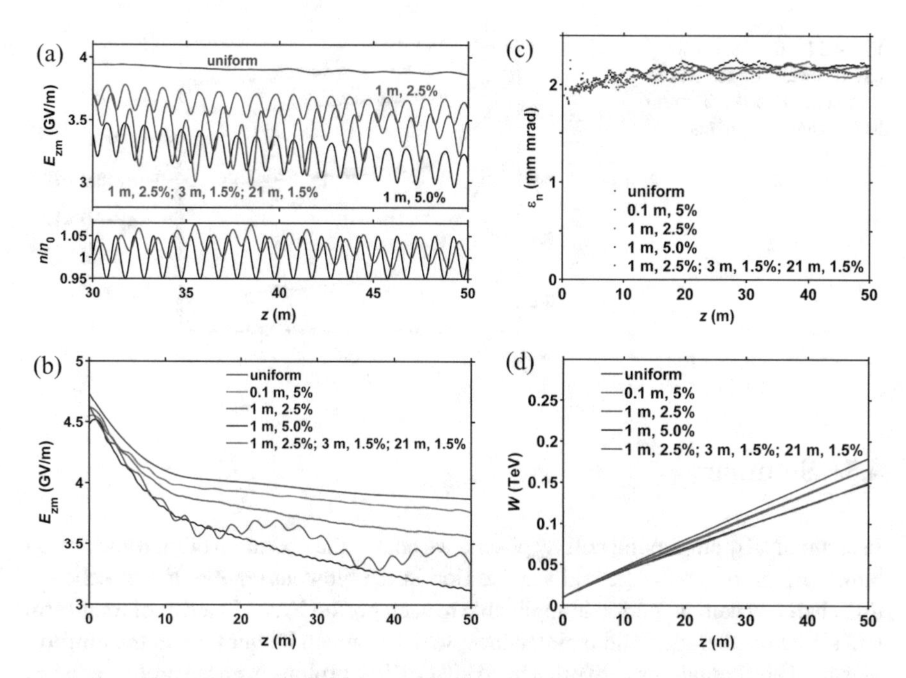

Fig. 4.10 **a** Maximum longitudinal electric fields excited in axially non-uniform plasmas versus the propagation distance. The field amplitudes (depicted in the upper plot) are smoothed in narrow intervals in z so that the oscillations are observable. The lower plot illustrates the corresponding plasma density profiles. The identically colored lines denote the same cases. **b** Maximum longitudinal electric field amplitudes are averaged in wide (1 m) intervals in z to facilitate the comparison of different cases of plasma inhomogeneity. **c** Normalized emittance of the witness beam in different cases of plasma inhomogeneity. **d** Mean energy of the witness beam versus the acceleration distance. Note that the "uniform" in the legends denotes the hollow channel with axially uniform plasma

4.4.3 Proton Survival Rate

In Sec. 4.3.2, we demonstrate that the survival rate of protons is essentially determined by the relation between their radial momenta and the potential well depth. Increasing the angular spread will cause more proton loss, as the protons with larger angular spread have enough transverse kinetic energy to escape from the potential well. Figure 4.11 gives a quantitative view of this. The normalized emittance of the protons in the simulated case (3.5 mm mrad, with the corresponding angular spread of 5×10^{-5}) is typical for the state-of-the-art beams [53, 54], thus there is no safety margin in beam emittance for our simulated case. Therefore, the multiple proton bunches employed in the discussed scheme are preferably generated by longitudinal modulation [22] instead of the radial self-modulation [14, 15] from a long proton bunch. The reason behind it is that the self-modulation gives rise to large radial momenta of the micro-bunches [20].

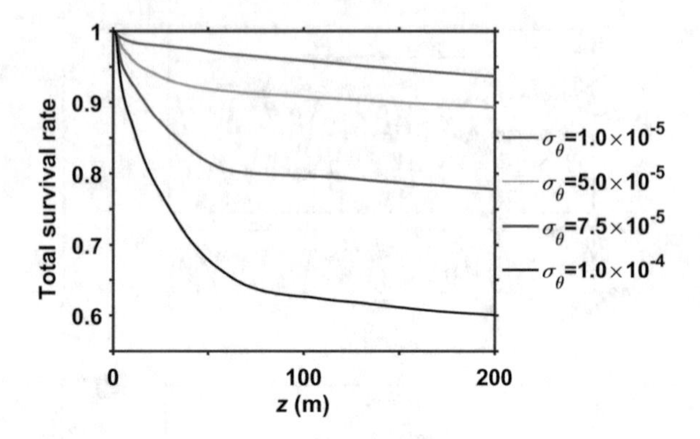

Fig. 4.11 Total survival rates concerning the whole proton driver with different initial angular spreads

4.5 Summary

In summary, by employing hollow plasma we enable the operation of multiple proton bunch driven plasma wakefield acceleration in the blowout regime. In this scheme, up to half a wakefield period is applicable to each proton bunch in terms of long-term and stable deceleration and maintenance, which is much longer than in the uniform plasma. The simulations verify that up to 90% of TeV protons from ten proton bunches survive after propagating through a plasma of length 150 m, bringing the witness electrons to 0.47 TeV with low energy spread (1.3%) and well-preserved normalized emittance. No strong dissipation of the accelerating mode is observed, which enables high efficiency of multi-bunch wave excitation. In addition, the hollow channel can sustain relatively large eigenfrequency fluctuations (at the level of 5%), which lowers the precision requirements in producing the desired hollow plasma channel. This work expands the concept of proton driven plasma wakefield acceleration and the hollow channels further open the path to emittance-preserving accelerating structures for future high-energy physics facilities.

References

1. Caldwell A, Lotov K, Pukhov A, Simon F (2009) Proton-driven plasma wakefield acceleration. Nat Phys 5(5):363
2. Lotov K (2010) Simulation of proton driven plasma wakefield acceleration. Phys Rev Spec Top-Accel Beams 13(4):041301
3. Yi L, Shen B, Lotov K, Ji L, Zhang X, Wang W, Zhao X, Yu Y, Xu J, Wang X et al (2013) Scheme for proton-driven plasma-wakefield acceleration of positively charged particles in a hollow plasma channel. Phys Rev Spec Top-Accel Beams 16(7):071301
4. Yi L, Shen B, Ji L, Lotov K, Sosedkin A, Wang W, Xu J, Shi Y, Zhang L, Xu Z et al (2014) Positron acceleration in a hollow plasma channel up to TeV regime. Sci Rep 4:4171

5. Li Y, Xia G, Lotov KV, Sosedkin AP, Hanahoe K, Mete-Apsimon O (2017) High-quality electron beam generation in a proton-driven hollow plasma wakefield accelerator. Phys Rev Accel Beams 20(10):101301
6. Zimmermann F (1997) A simulation study of electron-cloud instability and beam-induced multipacting in the LHC. Technical report, No. LHC-Project-Report-95
7. Mfietral E, Arduini G, Benedetto E, Burkhardt H, Shaposhnikova E, Rumolo G (2005) Transverse mode-coupling instability in the CERN super proton synchrotron. In: AIP conference proceedings, vol 773. AIP, pp 378–380
8. Mounet N (2012) The LHC transverse coupled-bunch instability. PhD thesis, Ecole Polytechnique Ffiedfierale de Lausanne, Lausanne, Switzerland
9. Assmann R, Papaphilippou Y, Giovannozzi M, Xia G, Zimmermann F, Caldwell A (2009) Generation of short proton bunches in the CERN accelerator complex. In: Proceedings of PAC09, Vancouver, BC, Canada, p 4542
10. Caldwell A, Lotov K, Pukhov A, Xia G (2010) Plasma wakefield excitation with a 24 GeV proton beam. Plasma Phys Control Fusion 53(1):014003
11. Xia G, Caldwell A, Lotov K, Pukhov A, Kumar N, An W, Lu W, Mori W, Joshi C, Huang C et al (2010) Update of proton driven plasma wake-field acceleration. In: AIP conference proceedings, vol 1299. AIP, pp 510–515
12. Xia G, Caldwell A (2010) Producing short proton bunch for driving plasma wakefield acceleration. In: Proceedings of IPAC, Kyoto, Japan, pp 4395–4397
13. Caldwell A, Lotov K (2011) Plasma wakefield acceleration with a modulated proton bunch. Phys Plasmas 18(10):103101
14. Lotov K (1998) Instability of long driving beams in plasma wakefield accelerators. In: Proceedings of the 6th European particle accelerator conference, Stockholm, 1998, pp 806–808
15. Kumar N, Pukhov A, Lotov K (2010) Self-modulation instability of a long proton bunch in plasmas. Phys Rev Lett 104(25):255003
16. Krall J, Joyce G (1995) Transverse equilibrium and stability of the primary beam in the plasma wake-field accelerator. Phys Plasmas 2(4):1326–1331
17. Whittum DH (1997) Transverse two-stream instability of a beam with a Bennett profile. Phys Plasmas 4(4):1154–1159
18. Lotov K (2015) Effect of beam emittance on self-modulation of long beams in plasma wakefield accelerators. Phys Plasmas 22(12):123107
19. Lotov K (2011) Controlled self-modulation of high energy beams in a plasma. Phys Plasmas 18(2):024501
20. Lotov K (2015) Physics of beam self-modulation in plasma wakefield accelerators. Phys Plasmas 22(10):103110
21. Caldwell A, Adli E, Amorim L, Apsimon R, Argyropoulos T, Assmann R, Bachmann A-M, Batsch F, Bauche J, Olsen VB et al (2016) Path to AWAKE: evolution of the concept. Nucl Instrum Methods Phys Res Sect A 829:3–16
22. Sheinman I, Petrenko A (2016) High-energy micro-buncher based on the mm-wavelength dielectric structure. In: 25th Russian particle accelerator conference (RuPAC'16), St. Petersburg, Russia
23. Lotov K (2013) Excitation of two-dimensional plasma wakefields by trains of equidistant particle bunches. Phys Plasmas 20(8):083119
24. Lotov K, Minakov V, Sosedkin A (2014) Parameter sensitivity of plasma wakefields driven by self-modulating proton beams. Phys Plasmas 21(8):083107
25. Lotov K (2017) Radial equilibrium of relativistic particle bunches in plasma wakefield accelerators. Phys Plasmas 24(2):023119
26. Katsouleas TC, Wilks S, Chen P, Dawson JM, Su JJ (1987) Beam loading in plasma accelerators. Part Accel 22:81–99
27. Chiou T, Katsouleas T, Decker C, Mori W, Wurtele J, Shvets G, Su J (1995) Laser wake-field acceleration and optical guiding in a hollow plasma channel. Phys Plasmas 2(1):310–318
28. Rosenzweig J, Breizman B, Katsouleas T, Su J (1991) Acceleration and focusing of electrons in two-dimensional nonlinear plasma wake fields. Phys Rev A 44(10):R6189

29. Pukhov A, Meyer-ter Vehn J (2002) Laser wake field acceleration: the highly non-linear broken-wave regime. Appl Phys B 74(4–5):355–361
30. Esarey E, Schroeder C, Leemans W (2009) Physics of laser-driven plasma based electron accelerators. Rev Mod Phys 81(3):1229
31. Hooker SM (2013) Developments in laser-driven plasma accelerators. Nat Photonics 7(10):775
32. Rosenzweig J, Cook A, Scott A, Thompson M, Yoder R (2005) Effects of ion motion in intense beam-driven plasma wakefield accelerators. Phys Rev Lett 95(19):195002
33. Kirby N, Berry M, Blumenfeld I, Hogan M, Ischebeck R, Siemann R (2007) Emittance growth from multiple Coulomb scattering in a plasma wakefield accelerator. In: IEEE particle accelerator conference (PAC). IEEE, pp 3097–3099
34. Mete O, Labiche M, Xia G, Hanahoe K (2015) GEANT4 simulations for beam emittance in a linear collider based on plasma wakefield acceleration. Phys Plasmas 22(8):083101
35. Corde S, Adli E, Allen J, An W, Clarke C, Clayton C, Delahaye J, Frederico J, Gessner S, Green S et al (2015) Multi-gigaelectronvolt acceleration of positrons in a self-loaded plasma wakefield. Nature 524(7566):442
36. Gessner S, Adli E, Allen JM, An W, Clarke CI, Clayton CE, Corde S, Delahaye J, Frederico J, Green SZ et al (2016) Demonstration of a positron beam-driven hollow channel plasma wakefield accelerator. Nat Commun 7:11785
37. Chiou T, Katsouleas T (1998) High beam quality and efficiency in plasma based accelerators. Phys Rev Lett 81(16):3411
38. Lee S, Katsouleas T, Hemker R, Dodd E, Mori W (2001) Plasma wakefield acceleration of a positron beam. Phys Rev E 64(4):045501
39. Kimura W, Milchberg H, Muggli P, Li X, Mori W (2011) Hollow plasma channel for positron plasma wakefield acceleration. Phys Rev Spec Top-Accel Beams 14(4):041301
40. Sahai AA, Katsouleas TC (2015) Optimal positron-beam excited plasma wakefields in hollow and ion-wake channels. In: Proceedings of IPAC15, Richmond, VA, USA, p 2674
41. Bulanov S, Kamenets F, Pegoraro F, Pukhov A (1994) Short, relativistically strong laser pulse in a narrow channel. Phys Lett A 195(1):84–89
42. Shvets G, Wurtele J, Chiou T, Katsouleas TC (1996) Excitation of accelerating wakefields in inhomogeneous plasmas. IEEE Trans Plasma Sci 24(2):351–362
43. Volfbeyn P, Lee P, Wurtele J, Leemans W, Shvets G (1997) Driving laser pulse evolution in a hollow channel laser wakefield accelerator. Phys Plasmas 4(9):3403–3410
44. Schroeder CB, Esarey E, Benedetti C, Leemans W (2013) Control of focusing forces and emittances in plasma-based accelerators using near hollow plasma channels. Phys Plasmas 20(8):080701
45. Schroeder C, Benedetti C, Esarey E, Leemans W (2013) Beam loading in a laser-plasma accelerator using a near-hollow plasma channel. Phys Plasmas 20(12):123115
46. Schroeder C, Benedetti C, Esarey E, Leemans W (2016) Laser-plasma based linear collider using hollow plasma channels. Nucl Instrum Methods Phys Res Sect A 829:113–116
47. Pukhov A, Jansen O, Tueckmantel T, Thomas J, Kostyukov IY (2014) Field-reversed bubble in deep plasma channels for high-quality electron acceleration. Phys Rev Lett 113(24):245003
48. Lotov K (2003) Fine wake field structure in the blowout regime of plasma wake-field accelerators. Phys Phys Rev Spec Top-Accel Beams 6(6):061301
49. Sosedkin A, Lotov K (2016) LCODE: a parallel quasistatic code for computationally heavy problems of plasma wakefield acceleration. Nucl Instrum Methods Phys Res Sect A 829:350–352
50. Ruth RD, Morton P, Wilson PB, Chao A (1984) A plasma wake field accelerator. Part Accel 17(SLAC-PUB-3374):171
51. Breizman B, Chebotaev P, Kudryavtsev A, Lotov K, Skrinsky A (1997) Self-focused particle beam drivers for plasma wakefield accelerators. In: AIP conference proceedings, vol 396. AIP, pp 75–88

52. Lotov K, Pukhov A, Caldwell A (2013) Effect of plasma inhomogeneity on plasma wakefield acceleration driven by long bunches. Phys Plasmas 20(1):013102
53. Gschwendtner E, Adli E, Amorim L, Apsimon R, Assmann R, Bachmann A-M, Batsch F, Bauche J, Olsen VB, Bernardini M et al (2016) AWAKE, the advanced proton driven plasma wakefield acceleration experiment at CERN. Nucl Instrum Methods Phys Res Sect A 829:76–82
54. Patrignani C, Particle Data Group et al (2016) Review of particle physics. Chin Phys C 40(10):100001

Chapter 5
High-Quality Positrons from a Multi-proton Bunch Driven Hollow Plasma Wakefield Accelerator

5.1 Introduction

In Chap. 4, we have demonstrated that, by means of a hollow plasma, multiple proton bunches work well in driving nonlinear plasma wakefields and accelerate electrons to the energy frontier with preserved beam quality. For the future energy frontier lepton colliders, the generation of TeV-scale positrons is equally important. However, the acceleration of positrons is different as positrons see a different and less favorable accelerating structure, which is strongly charge dependent. The consequence is a discrepancy in terms of the beam quality, that is, either a preserved normalized emittance with a high energy spread or vice versa. This results from the conflict that the plasma electrons, used to provide focusing to the multiple proton bunches, now dilute the positron bunch. In the following we first explain this issue. Then we propose two intuitive approaches which are capable of mitigating the beam quality deterioration yet to a limited extent. Fortunately, afterwards we validate that a combined solution is able to guarantee both high energy gain and high beam quality. Finally, we analyze the dependent relations between the acceleration and beam and plasma parameters.

5.2 Challenges in Positron Acceleration

Plasma wakefield accelerators have become increasingly popular in constructing compact and cost-effective colliders in the energy level of hundreds of GeV or TeV [1–3] owing to the ultra-high accelerating gradient. There has been essential progress in plasma-based electron acceleration to high energy [4] or with high efficiency [5], but little advance for positron acceleration. The most fundamental obstacle is the lack of a stable acceleration regime for positrons. As seen from Fig. 5.1, in the blowout regime of plasma wakefield acceleration, the plasma electrons are displaced off axis by the electron driver and then concentrate in the bubble sheath. The entire rear half of

Y. Li, *Studies of Proton Driven Plasma Wakefield Acceleration*, Springer Theses,
https://doi.org/10.1007/978-3-030-50116-7_5

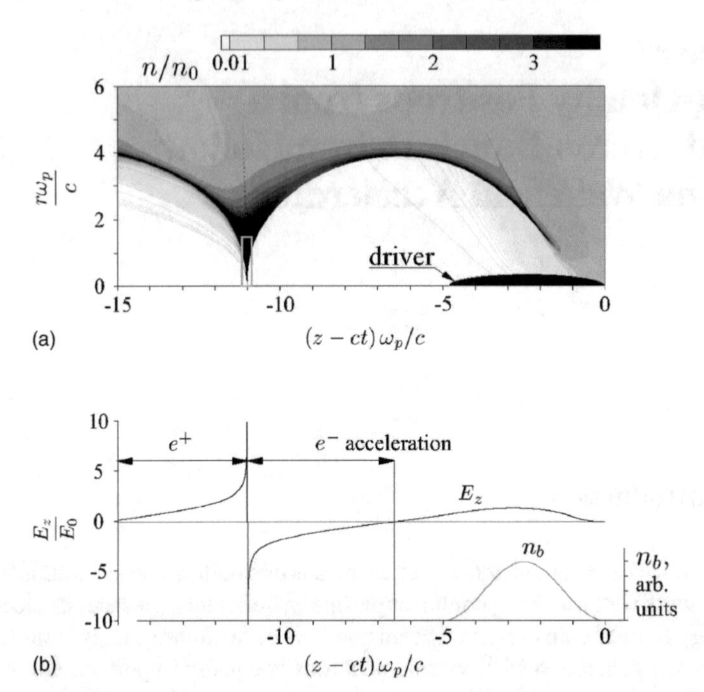

Fig. 5.1 **a** Bubble shape driven by an intense electron beam; **b** Lineouts of the on-axis excited electric field and the driver density. Images reproduced from reference [8]

the bubble which is devoid of plasma electrons forms an ideal accelerating structure for the electron beam, i.e., is both uniformly accelerating and linearly focusing along the beam radius [6, 7]. However, the focusing region for positrons is substantially smaller being at the back of the bubble (the region marked by the frame in Fig. 5.1a), where the plasma electrons collapse on axis and the plasma density is nonuniform. Therefore, the acceleration and focusing for positrons vary radially, which leads to the growth of the energy spread and the emittance. Also it requires precise control of the location of positrons [8]. In consequence, the idea in Ref. [8] of accelerating positrons in the nonlinear wake of the electron beam is apparently challenging [9]. Nevertheless, an encouraging experiment in 2015 [10] validated that it is feasible to accelerate positrons at the tail by the front of the positron beam. By sufficiently self-loading the wake, the rear of the beam sees focusing of the "sucked-in" plasma electrons and a flat accelerating field. Despite a low energy spread, the preservation of the beam emittance is still an issue [11, 12].

5.3 Applications of Hollow Plasma in Positron Acceleration

A hollow channel plasma was initially proposed to confine the lasers [13]. Later it was recognized to strongly benefit the acceleration and quality of positrons [14, 15]. Take the work in Ref. [14] for example, the hollow plasma dispels the defocusing from the background ions and also nonlinear focusing from the plasma electrons attracted towards the axis. In addition, the longitudinal wakefields become uniform transversely. The limitation is here the hollow channel acts as a waveguide structure. It requires no plasma electrons streaming into the channel and hence no intense drivers, which are supposed to be used for large wakefield excitation. Reference [15] elucidates strongly nonlinear beam-plasma interactions, whereas the ultra-short and dense proton driver is hardly obtainable, because the real proton bunch in practice is tens of cm long.

5.4 Simulation Results

In Chap. 4, we exploit multiple equidistant proton bunches, which are hypothetically obtained from the longitudinal modulation of a long bunch. This eases the challenge in compressing it to one single short proton bunch. Furthermore, the introduction of the hollow channel removes the ion defocusing to the proton bunches. It also overcomes deterioration of the witness electron beam quality in uniform plasma, which results from the asymmetric plasma response to the positively charged drivers compared to their counterparts. As a consequence, the multiple proton bunches work well in the nonlinear regime sourcing strong plasma fields and accelerating electrons to the energy frontier with preserved beam quality.

In this chapter, likewise we adopt multiple proton bunches as the driver along with the indispensable part—a hollow plasma channel for the nonlinear wake excitation and validate the efficient acceleration and beam quality preservation of the positron bunch.

5.4.1 Dilemma in Preserving the Beam Quality

Table 5.1 displays the simulation parameters. The proton beam and plasma parameters are taken based on the same rules demonstrated in Chap. 4. To be specific, the multiple equidistant proton bunches are presumably obtained from a single LHC bunch by longitudinal bunch modulation, hence the population of each proton bunch is the population of a LHC bunch divided by the number of multiple proton bunches. The bunch energy spread is set as 10%, larger than that of the LHC bunch as we assume a longitudinal bunch compression of the LHC bunch before modulated to accommodate the need for shorter proton bunches. The bunches are supposed to

Table 5.1 Simulation parameters

Parameters	Values	Units
Initial proton driver		
Single bunch population	1.28×10^{10}	
Energy	1	TeV
Energy spread	10%	
Single bunch length	66	μm
Single bunch radius	68	μm
Bunch train period	660	μm
Initial witness positron bunch		
Population, N_p	1.0×10^9	
Energy, W_0	10	GeV
Energy spread, $\delta W/W$	1%	
RMS length, σ_z	5	μm
RMS radius, σ_r	4	μm
Normalized emittance, ϵ_n	0.25	μm
Hollow plasma channel		
Plasma density, n_0	5×10^{15}	cm^{-3}
Channel radius, r_c	190	μm
Quadrupole magnets		
Magnetic field gradient, S	500	T/m
Quadrupole period, L	0.9	m

be in tune with the wakefields, the period of which is determined by the plasma density and the channel radius. An optimal combination of the plasma density and the channel radius occurs when the proton bunches drive strong and stable wakefields. Quadrupole magnets are used to prevent the head of the first proton bunch from emittance-driven erosion owing to no or weak plasma focusing therein, and also confine the positron bunch within a small radius. In theory, the stronger the quadrupoles, the better the focusing. Here we keep them in an acceptable level considering current technologies. More importantly, stronger quadrupoles require the initial positron energy to be larger, otherwise the positron bunch will diverge under the quadrupole focusing. We have discussed the constraints in Chap. 3.

The proton beam dynamics and the wake characteristics are similar to those in Chap. 4 as well and thus not repeated here. Yet it is worth mentioning that the proton driver and plasma channel have been carefully optimized in order to radially enlarge the acceleration region for positrons which is free from plasma electrons. This is essential because, as will be seen later, the positrons interfere severely with plasma electrons penetrating near axis. Given the same reason, the initial radius and emittance of the positron bunch are required to be small. The positron bunch is initialized with the equilibrium radius, which is calculated based on Ref. [16] assuming positrons are only focused by the external quadrupole fields. The quasi-static particle-in-cell code LCODE based on the 2D cylindrical geometry has been employed due to its high computing efficiency to conduct all the simulations, which extend over long distances whilst maintaining a simulation grid fine enough to resolve the positron beam.

Figure 5.2a shows the axisymmetric space distribution of the plasma electrons (2D map) and proton (grey points) and positron (red points) bunches in a co-moving window where the position coordinate $\xi = z - ct$ is used for convenience. To facilitate comparisons, ξ is normalized to the plasma skin depth c/ω_p, where c is the light velocity and ω_p is the plasma electron frequency corresponding to the initial plasma density of n_0. There are no plasma ions within the hollow channel. Each proton bunch basically stays in the rear half of the bubble seeing decelerating fields (Fig. 5.2b). The positron bunch is located within the first half of the bubble just behind the last proton bunch (i.e., the 9th bunch), where the longitudinal field is accelerating and radially uniform. In the initial acceleration stage (before $z = 84$ m), there are no plasma electrons penetrating into the positron region, which makes the positrons only focused by the quadrupole fields. With its initial radius matching with the focusing structure, the positron bunch is stably accelerated with a well-preserved normalized emittance.

Unfortunately, the slope of the accelerating field initially seen by positrons is positive and is not conducive to beam loading. This causes an increasing energy spread due to head positrons experiencing larger fields than the tail (see Fig. 5.4 later). This type of region where the bunch is accelerated with the energy spread increasing is termed as "$\mathrm{I_{acc}}$". As protons have larger mass, their relativistic factors are essentially smaller in comparison with the ultra-relativistic positrons being further accelerated. Therefore, the protons and the wake phase keep shifting backwards with respect to the positrons. This creates a perfect opportunity for the positron bunch to get into the "$\mathrm{D_{acc}}$" region ahead of the maximum accelerating field. The field here has a sharp and negative slope and could conceivably rectify the energy spread. However, this region resides in the bubble head, where the plasma electrons stream across the axis and the plasma density is uneven. The radial fields thus vary transversely, which is detrimental to the beam emittance.

Reference [15] has still taken advantage of the "$\mathrm{D_{acc}}$" region. However, in this previous study, the single short proton driver is required to be dense enough so that an elliptical bubble nearly devoid of plasma electrons is formed behind the driver under strongly nonlinear wake excitation. That is, the perturbed plasma electrons move within a thin bubble sheath. In this way, a small radial space free from plasma electrons at the bubble head is maintained for the positron bunch.

Nevertheless, it is not the case in our configuration. The multiple proton bunches will lose focusing from the plasma electrons and diverge significantly [17]. This in turn leads to more plasma electrons flowing in and ruining the positron beam quality. Furthermore, with proton energy depletion and the bunch elongation after a long distance, it is increasingly tricky to keep the plasma electrons close to the bubble boundary. Figure 5.2c foresees the positron emittance growth from $z = 84$ m, when some plasma electrons get closer to the axis and the radial "clean" space for positrons becomes smaller. The plasma electron trajectories crossing with that of

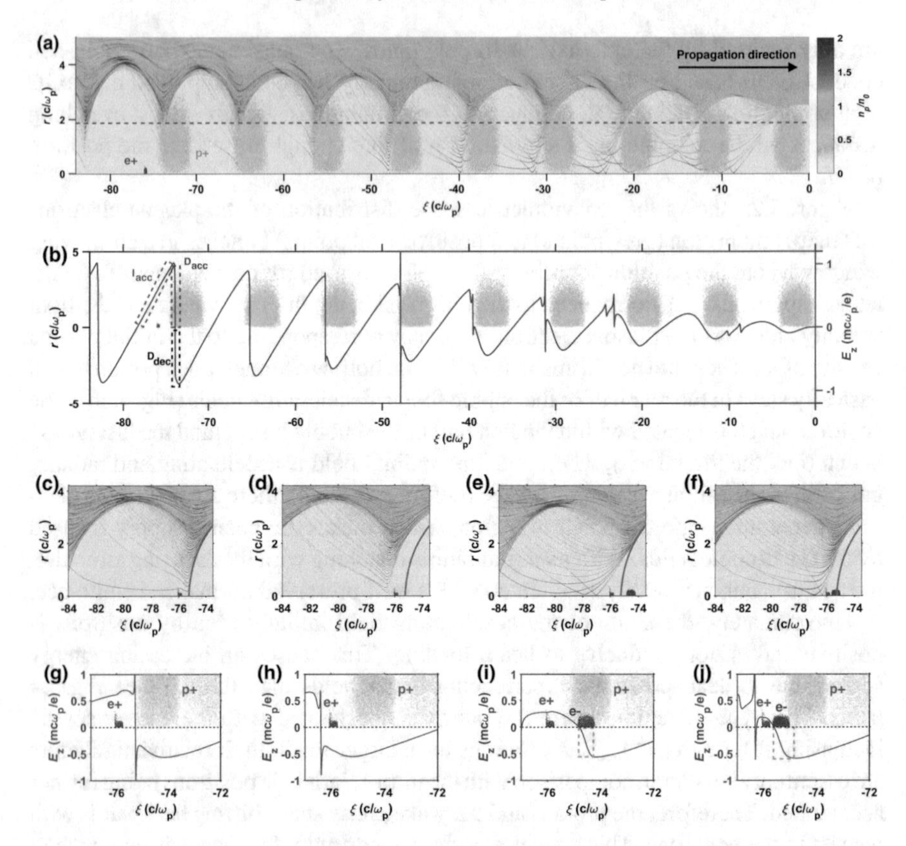

Fig. 5.2 Axisymmetrical spatial distribution of the plasma electrons, proton and positron bunches (a) and the corresponding on-axis longitudinal electric field (b) at $z = 0$ m, the close-ups of plasma electron distribution (the 3rd row) and the on-axis longitudinal electric field (the 4th row) in the vicinity of the positron bunch at $z = 84$ m for the cases of the original configuration (c, g), decreased by 9% plasma density (d, h), loading of extra electrons (e, i), and decreased plasma density with electron presence (f, j), respectively. The grey points denote the protons while the red and blue points represent the positrons and extra electrons respectively in all sub-figures. The value of the color scale denoting the perturbed plasma density normalized to the initial plasma density in **a** is limited to 2 for the trajectories of the plasma electrons to stand out. The dashed red line in **a** outlines the hollow channel boundary. The red, green and black frames in **b** mark the “I_{acc}”, “D_{acc}” and ‘D_{dec}” regions for positrons, respectively. The plots in the third row share the same color scale as that in **a** and thus not repeated. The magenta curves sketch the boundary of the bubble ahead of the positron bunch under the original configuration. The orange frames in **i** and **j** mark the “A_{acc}” region for electrons

positrons dilute the positron bunch. However, by then the bunch has not reached the “D_{acc}” region (Fig. 5.2g). Overall, there is a discrepancy between obtaining a small beam emittance and a small energy spread in our studied case. Basically the positron bunch needs to get to the bubble head to lower its energy spread gained before. Nevertheless, before that the nonuniform plasma electrons will destroy its emittance remarkably. As a result, either the acceleration is truncated before the beam emittance gets ruined or some measures must be taken from this point.

5.4.2 *Solution 1: Plasma Density Decrease*

Recall that the "D_{acc}" region is conducive to the decrease of the energy spread due to the negative field slope. It follows that the "D_{dec}" region should contribute to reducing the energy spread as well. Despite being decelerating for positrons, this region is free from plasma electrons in a large radial space as it is located at the rear of the bubble ahead of the positron bunch. Thus, it is expected to be capable of conserving the beam emittance while reducing its energy spread. Since the bunch will deteriorate if it continuously slides into the "D_{dec}" region by means of phase dephasing, we propose to reduce the plasma density to $0.91n_0$ from the point when the beam emittance starts to degrade. In this way, the wake wavelength increases and the bubble shifts backwards (Fig. 5.2d). It looks like the positron bunch directly jumps from the front of the old bubble to the tail of the bubble ahead (i.e., from Fig. 5.2c to Fig. 5.2d).

It is noteworthy that, while the plasma density change brings substantial phase shift in terms of the positrons, its effect on the resonance of proton bunches is acceptable. To be specific, the plasma wavelength λ is inversely proportional to the square root of the plasma density n, therefore the slight density decrease $\delta n/n_0$ will cause the plasma wavelength to increase by $\delta\lambda/\lambda_0 = 0.5\delta n/n_0$ where λ_0 is the initial plasma wavelength. It follows that with a relative density decrease of 9%, the wake period increases by 4.5%. This implies a negligible phase shift with respect to the first proton bunch considering the almost half wavelength it occupies. However, the phase shift accumulates from the first to the last bunch and it adds up to about 0.72 plasma wavelength relative to the last bunch. As initially the last bunch lags behind the maximum decelerating field (Fig. 5.2b), the wake shift affects it insignificantly and most of the protons still reside in the deceleration region.

Note that there is already some dilution to the positron bunch in the large radius at the time of plasma density change. This is done deliberately to keep the bunch longer in the initial acceleration stage with a large gradient, and it is acceptable since the core beam (95% of positrons) is still conserved (Fig. 5.3). Figure 5.4 (solution 1) illustrates the evolution of the energy gain and energy spread of the core positron bunch in the initial acceleration stage (before $z = 84$ m) and the stage devoted to reducing the energy spread (after). We see that the decrease of the energy spread is essentially inefficient and even goes to the opposite way after a short distance when the bunch slips out of the "D_{dec}" region. Also it comes with a large energy gain loss. The reasons are threefold. First of all, the plasma density change, although small, leads to less resonance of the proton bunches. Secondly, the proton bunches deform significantly after long term depletion. These result in weaker wake excitation and hence less sharp field slope compared to the initial stage (Fig. 5.2b). Thirdly, the

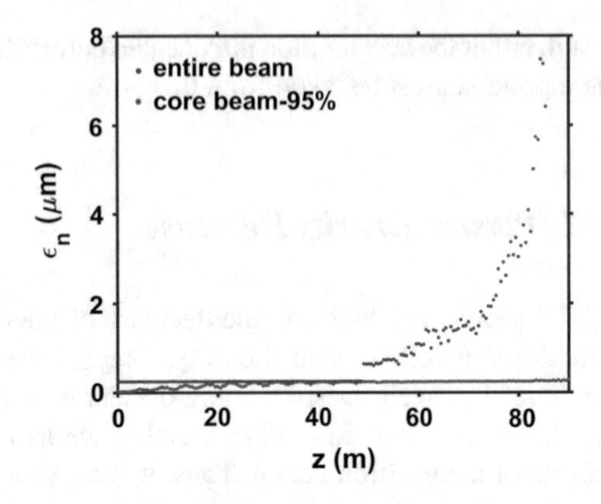

Fig. 5.3 Evolution of the normalized emittance of the positron bunch with the acceleration distance. It shows the results for the entire positron beam (blue) and the core beam (red), where the radial positions of positrons are within $0.1c/\omega_p$ and the angular spread is smaller than 10^{-6} rad

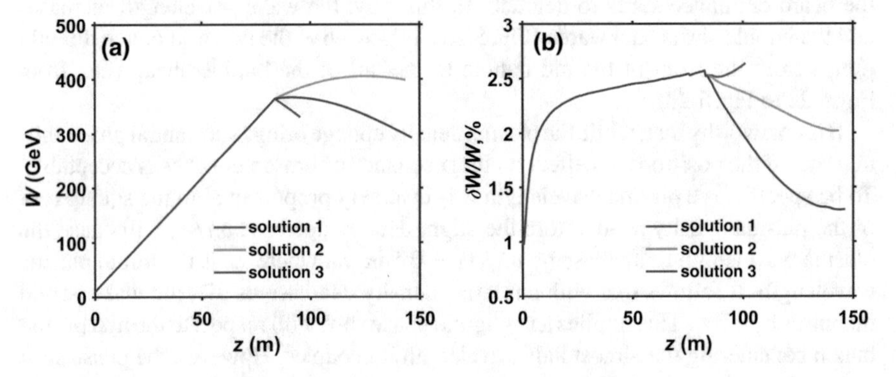

Fig. 5.4 Evolution of the positron energy gain (a) and energy spread (b) for the three proposed solutions starting at $z = 84$ m to reduce the energy spread while preserving the beam emittance. Here it only shows the results for the core positron beam

wake flattens more under the beam loading and the extension of the wake period (Fig. 5.2h). Regardless, the positron beam emittance is still preserved.

5.4.3 Solution 2: Presence of the Electron Beam Load

Since the plasma electrons are the main obstacle to the emittance preservation of the positrons, we introduce an extra electron bunch to the bubble tail ahead of the positron bunch (Fig. 5.2e). The electron bunch has a large population of 2×10^{10}. Its radial space charge force acts back on the bubble sheath electrons and delays them from returning to the ξ-axis. Therefore, the plasma electrons are expelled away from the positrons and the bubble shape is altered (Fig. 5.2e). Figure 5.2i shows that the loading of electrons changes the wake form as well and the positron bunch sees a flat or slightly increasing wakefield at the beginning. The negative slope ahead is expected to shift towards the positrons after a distance. Note that the electron bunch cannot be loaded to the decelerating field region as the energy loss will lead to its quick divergence under the radial focusing of quadrupoles [16]. Figure 5.4 indicates that loading of electrons (solution 2) reduces the energy spread much more efficiently compared to the plasma density change (solution 1). Another positive feature is that the positron energy increases continuously with preserved beam emittance until it gets further into the "D_{dec}" region.

5.4.4 Solution 3: Combination

The shortcoming of the second solution is when loading the electrons, the positrons are still far away from the "D_{acc}" region and the fields seen are not effective to reduce the energy spread. When the positron bunch arrives at the "D_{acc}" region, the field slope by then becomes insignificant due to large energy loss to both positrons and electrons and wake decrease. Hence, we combine the aforementioned two approaches and propose to load extra electrons while reducing the plasma density to $0.95n_0$. In so doing, the positron bunch directly jumps to the "D_{acc}" region which has a large slope (Fig. 5.2j) and meanwhile the presence of electrons removes the interference of plasma electrons with positrons (Fig. 5.2f). More importantly, the required electron population is 1×10^{10} to maintain enough "clean" space for positrons. This is twice smaller than in solution 2 with only electron loading, where the electrons are further away from the positrons due to the requirement of staying in the accelerating field region and thus it requires a larger number of electrons to bend the bubble shape more. In addition, the plasma density decrease is less than in solution 1, as it only needs to shift the "D_{acc}" region towards the positron bunch instead of the "D_{dec}" region further ahead. As a result, more protons stay within the decelerating region.

Figure 5.4 (solution 3) demonstrates the successful decrease of the energy spread to 1.3%. It also happens substantially quicker than the other two solutions. Over 95% of positrons are kept at the initial normalized emittance (0.25 mm mrad). The net energy gain in the stage of reducing the energy spread is negative, which is inferior to solution 2, where the dephasing length sees accelerating fields longer. For all solutions, the reduction of energy spread slows down when the positron bunch approaches the maximum decelerating field in the "D_{dec}" region, whose slope is trivial or almost zero. The end of this stage corresponds to the bunch getting out of the "D_{dec}" region, otherwise the energy spread will increase again.

While accelerating the positrons, the electron bunch is accelerated from 10 GeV to 150 GeV as well but its energy spread is as large as 17% because it mostly stays in the "A_{acc}" region, where the field is accelerating but the slope is not conducive. Its normalized emittance is well conserved in the whole process, as no plasma focusing acts on it. The further energy extraction by electrons suggests an increase of the overall energy efficiency for this particular configuration if this electron bunch is applicable.

By comparison, we see the most effective way to reduce the energy spread while preserving the beam emittance is to load an extra electron bunch and reduce the plasma density from the point when the plasma electrons start to interfere with the positrons. Despite that, the first two proposed methods are useful when extending to other set-ups. For example, the idea of plasma density change could be introduced into the single electron/positron/laser beam driven positron acceleration cases in hollow plasma, where the acceleration distance is not long and the wake form does not change significantly. In addition, the idea of an additional electron bunch is expected to be beneficial to the AWAKE scheme (details will be introduced in Chap. 7). The load of an extra electron bunch can expel the nonuniform plasma electrons away from the witness electron bunch, so that it can be accelerated with high quality. A different use can be found in Ref. [18], where an additional tailored escort electron beam with a very high charge was introduced into a later phase of the electron acceleration. It overlapped with the witness electron bunch and overloaded the plasma wave, reversing the local accelerating field slope. In this way, the witness energy spread was decreased by an order of magnitude.

5.5 Acceleration Characteristics

5.5.1 Dependence on the Electron Population and Plasma Density

In this section, we further discuss the dependence of acceleration performance and beam quality preservation on plasma and the electron bunch in terms of the third solution discussed in Sect. 5.4.4. The electron bunch is restricted to the following conditions. First of all, the electron charge is enough to expel plasma electrons away from the positron bunch. Second, the electron bunch resides in the acceleration region and better exactly ahead of the zero field [like the electron bunches in Fig. 5.2i, j], so that the energy extraction from protons is the least.

Given initially the same decreased plasma density of $0.95n_0$, Fig. 5.5 compares three cases under different populations of electrons. The electron bunch with larger charge resides further away from the positron bunch as more electrons overload the wake field to zero earlier. Therefore, it leaves a longer distance between the positron bunch and the zero field (Fig. 5.5a), i.e., a longer dephasing length for acceleration before positrons getting into the decelerating field region. As a result, the net energy gain of positrons for the larger electron loading case is higher at the same distance (Fig. 5.5b). Nevertheless, the wake field slope seen by the positron bunch is smaller and thus the energy spread drops slower (Fig. 5.5c).

It has been demonstrated that a plasma density drop to $0.95n_0$ shifts the "D_{acc}" region towards the positrons. Although with a larger plasma density the effect of

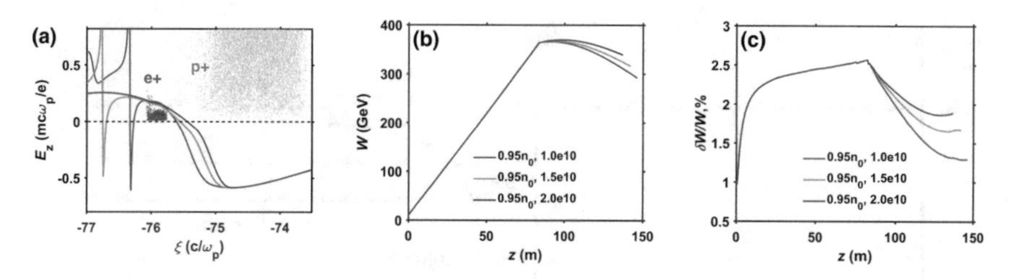

Fig. 5.5 The on-axis longitudinal electric field seen by the positron bunch at $z = 84$ m when starting loading the extra electron bunch (a) and evolution of energy gain (b) and energy spread (c) for three cases of different electron charge loading under the same decreased plasma density of $0.95n_0$. The red and grey points denote positrons and protons respectively. The electron bunch is not marked here, but it should be easy to figure out its position as its leftmost always resides at the zero field for each case

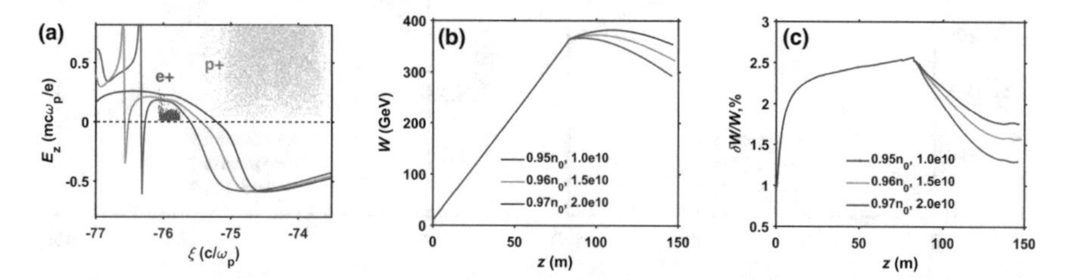

Fig. 5.6 The on-axis longitudinal electric field seen by the positron bunch at $z = 84$ m when starting loading the extra electron bunch (a) and evolution of energy gain (a) and energy spread (b) for three cases of different electron charge loading accompanied by different decreased plasma densities. The red and grey points denote positrons and protons respectively. The electron bunch is not marked here

phase shift on multiple proton bunches will be smaller, because the electron bunch is further away from the positron bunch, it requires a larger electron loading so that the space charge force is enough to repel the plasma electrons more. Figure 5.6 shows the cases where the larger dropped plasma density comes with a larger electron load. Similar to the case in Fig. 5.5, the energy spread decreases slower with a large electron load, but the net energy gain increases. Also when comparing Figs. 5.5 and 5.6, it is apparent that with the same electron loading, the wakefield and energy gain are promoted under a larger dropped plasma density. This is because more protons contribute to the wake excitation under less phase shift.

5.5.2 *Different Longitudinal Plasma Profiles*

In Sect. 5.4.4, we see the net energy gain in the third solution (over the stage of reducing energy spread) is negative because of positrons later sliding into the "D_{dec}" region with wake phase shift. Then it is natural to think of further increasing the

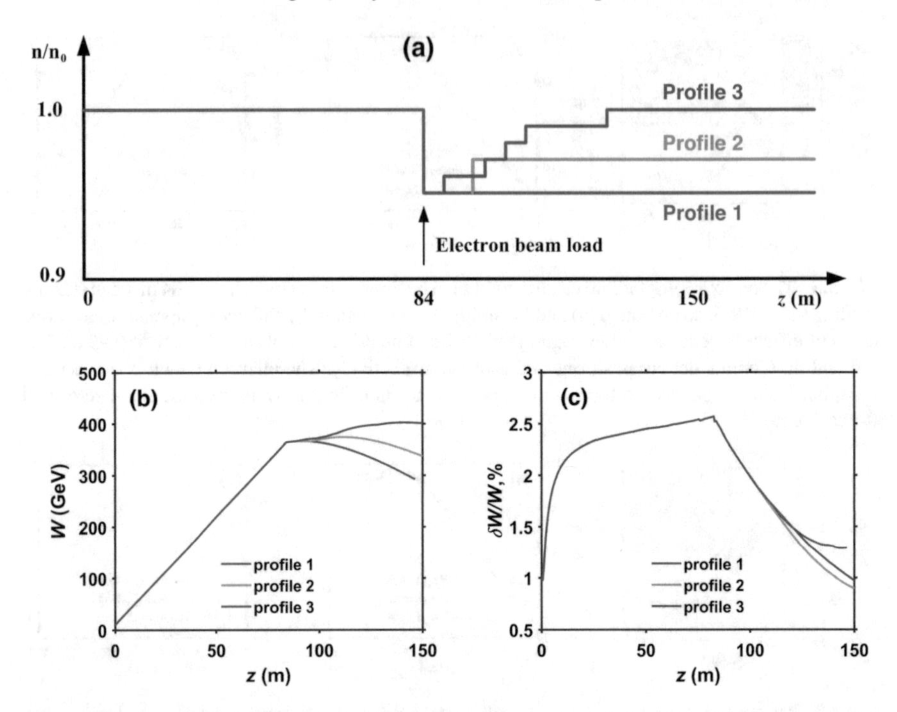

Fig. 5.7 Three different types of longitudinal plasma profiles (a) and the corresponding evolution of energy gain (b) and energy spread (c) of the positrons being accelerated in such plasma channels. The identical colour represents the same case. The red one corresponds to the third solution presented in Sect. 5.4.4

plasma density, so that the wake wavelength decreases and the "D_{acc}" region is shifted back to the positron bunch. Meanwhile, with a wake wavelength decrease, the extra electron bunch is expected to move towards a smaller acceleration gradient and extract less energy from the plasma wave.

Figure 5.7 compares the cases with different longitudinal plasma profiles adopted along with the same electron beam loading (1×10^{10}). The red coloured case represents the third solution discussed in Sect. 5.4.4. With a slight plasma density increase from $0.95n_0$ to $0.96n_0$ (i.e., the green line in Fig. 5.7a), both the net energy gain and the decrease rate of energy spread are promoted [Fig. 5.7b, c]. Still, for a longer acceleration distance, the positrons will inevitably get into the "D_{dec}" region again. Therefore, a more complicated plasma profile (i.e., the blue line in Fig. 5.7a) is proposed where the plasma density increase occurs when or after the positrons slip into the "D_{dec}" region. In this case, the positron energy increases continuously to 400 GeV, while the energy spread decreases to the initial level of 1% or even less with a longer distance at almost the same rate. Here the plasma with a stepped density can be realized by segmented plasma in practice. In addition, it allows a substantial response distance to change the plasma density, as the wake phase changes insignificantly in terms of the acceleration distance. For instance, the wake only dephases by around

$8.4 \times 10^{-3} c/\omega_p$ if passing through a 1 m longer plasma. Further simulations prove that the density change is not necessarily sharp and a gradual change even helps the acceleration performance.

5.6 Summary

In this chapter, we first analyse the issues in preserving the quality of the positron bunch while accelerating it to high energy in the multiple proton bunch driven hollow plasma wakefield accelerator. It is difficult to maintain a small normalized emittance and a small energy spread simultaneously. This results from the plasma electrons which, although providing focusing to the multiple proton bunches, dilute the positron bunch. We propose and compare three different solutions, and find that the most efficient way is to load extra electrons and meanwhile decrease the plasma density from the point when the positron emittance starts to degrade. Simulations show that the positron bunch can be accelerated to 40% of the driver energy with the energy spread as low as 1% when adopting a slightly more sophisticated plasma profile. Its normalized emittance is well preserved at the initial value. The electron beam charge must match the plasma density decrease to reach the best acceleration condition. For a smaller density decrease, the electron beam loading must be stronger.

This work expands the concept of positron acceleration driven by protons. The obtainable high quality and high energy positrons are promising candidates for the future energy frontier lepton colliders. In addition, the proposed ideas are not restricted to the scheme presented here and they should find applications in other set-ups.

References

1. Lee S, Katsouleas T, Muggli P, Mori W, Joshi C, Hemker R, Dodd E, Clayton C, Marsh K, Blue B et al (2002) Energy doubler for a linear collider. Phys Rev Spec Top-Accel Beams 5(1):011001
2. Adli E, Muggli P (2016) Proton-beam-driven plasma acceleration. Rev Accel Sci Technol 9:85–104
3. Hogan MJ (2016) Electron and positron beam driven plasma acceleration. Rev Accel Sci Technol 9:63–83
4. Blumenfeld I, Clayton CE, Decker F-J, Hogan MJ, Huang C, Ischebeck R, Iverson R, Joshi C, Katsouleas T, Kirby N et al (2007) Energy doubling of 42 GeV electrons in a metre-scale plasma wakefield accelerator. Nature 445(7129):741
5. Litos M, Adli E, An W, Clarke C, Clayton C, Corde S, Delahaye J, England R, Fisher A, Frederico J et al (2014) High-efficiency acceleration of an electron beam in a plasma wake field accelerator. Nature 515(7525):92
6. Rosenzweig J, Breizman B, Katsouleas T, Su J (1991) Acceleration and focusing of electrons in two-dimensional nonlinear plasma wake fields. Phys Rev A 44(10):R6189
7. Lu W, Huang C, Zhou M, Tzoufras M, Tsung F, Mori W, Katsouleas T (2006) A nonlinear theory for multidimensional relativistic plasma wave wakefields. Phys Plasmas 13(5):056709

8. Lotov K (2007) Acceleration of positrons by electron beam-driven wakefields in a plasma. Phys Plasmas 14(2):023101
9. Wang X, Muggli P, Katsouleas T, Joshi C, Mori W, Ischebeck R, Hogan M (2009) Optimization of positron trapping and acceleration in an electron beam-driven plasma wakefield accelerator. Phys Rev Spec Top-Accel Beams 12(5):051303
10. Corde S, Adli E, Allen J, An W, Clarke C, Clayton C, Delahaye J, Frederico J, Gessner S, Green S et al (2015) Multi-gigaelectronvolt acceleration of positrons in a self-loaded plasma wakefield. Nature 524(7566):442
11. Hogan M, Clayton C, Huang C, Muggli P, Wang S, Blue B, Walz D, Marsh K, O'Connell C, Lee S et al (2003) Ultrarelativistic-positron-beam transport through meter-scale plasmas. Phys Rev Lett 90(20):205002
12. Muggli P, Blue B, Clayton C, Decker F, Hogan M, Huang C, Joshi C, Katsouleas TC, Lu W, Mori W et al (2008) Halo formation and emittance growth of positron beams in plasmas. Phys Rev Lett 101(5):055001
13. Katsouleas T, Chiou T, Decker C, Mori W, Wurtele J, Shvets G, Su J (1992) Laser wakefield acceleration & optical guiding in a hollow plasma channel. In: Proceedings of AIP conference, vol 279. AIP, pp 480 489
14. Gessner S, Adli E, Allen JM, An W, Clarke CI, Clayton CE, Corde S, Delahaye J, Frederico J, Green SZ et al (2016) Demonstration of a positron beam-driven hollow channel plasma wakefield accelerator. Nat Commun 7:11785
15. Yi L, Shen B, Ji L, Lotov K, Sosedkin A, Wang W, Xu J, Shi Y, Zhang L, Xu Z et al (2014) Positron acceleration in a hollow plasma channel up to TeV regime. Sci Rep 4:4171
16. Lotov K (2010) Simulation of proton driven plasma wakefield acceleration. Phys Rev Spec Top-Accel Beams 13(4):041301
17. Li Y, Xia G, Lotov KV, Sosedkin AP, Hanahoe K, Mete-Apsimon O (2017) Multi-proton bunch driven hollow plasma wakefield acceleration in the nonlinear regime. Phys Plasmas 24(10):103114
18. Manahan GG, Habib A, Scherkl P, Delinikolas P, Beaton A, Knetsch A, Karger O, Wittig G, Heinemann T, Sheng Z et al (2017) Single-stage plasma based correlated energy spread compensation for ultrahigh 6d brightness electron beams. Nat Commun 8:15705

Chapter 6
Assessment of Misalignment Induced Effects in Proton Driven Hollow Plasma Wakefield Acceleration

6.1 Introduction

Hollow plasma channel has played a key role in our former studies of proton-driven plasma wakefield acceleration. It overcomes the issue of beam quality degradation caused by the nonlinear transverse wakefields varying in radius and time in uniform plasma [1]; enables the stable driving characteristics of multiple proton bunches in the nonlinear regime [2]; and provides a quality-beneficial and high gradient accelerating structure for the positron beam [3]. Owing to these positive features, both the electron and positron beams have been demonstrated in simulations to be accelerated to the energy frontier with well-preserved beam quality in a long hollow plasma channel, which advances the development of future TeV-scale lepton colliders.

On the other hand, the proposed hollow channel scheme imposes a stringent requirement on the beam-channel alignment. This is difficult in reality as random laser jitters or other unexpected factors result in offset of either the channel or the beam. Asymmetric and possibly strong transverse wakefields are thus induced throughout the channel by the misaligned beam, which could deflect the beam, leading to a significant beam loss. They also very likely jeopardise the beam quality. All these make the beam unsuitable for the applications on the colliders, as the resultant luminosity is reduced appreciably. In this chapter, we demonstrate the beam and wakefield characteristics under the misalignment and propose and assess solutions to mitigate the adverse transverse effects.

6.2 Beam-Channel Misalignment in the Weakly Linear Regime

Reference [4] derives analytically the beam-channel misalignment where the hollow channel acts as a waveguide, and describes the beam-breakup (BBU) instability [5] driven by the induced transverse wakefield associated with the dipole mode. In

Y. Li, *Studies of Proton Driven Plasma Wakefield Acceleration*, Springer Theses,
https://doi.org/10.1007/978-3-030-50116-7_6

general, the off-axis or asymmetric beam excites the dipole mode whose strength increases linearly with the beam offset Δr and strongly depends on the channel radius. The transverse force shares the same direction as the beam offset and thus pushes the beam further away from the axis. As the deflection force grows in strength along the bunch, the trailing particles behind the bunch head are driven even further, exciting the mode more. As a result, the tail particles see the largest transverse force and are rapidly deflected towards the channel wall. This process results in a banana shape of the bunch, which is deleterious to the beam emittance. In addition, the growth length of the BBU instability scales roughly linearly with the channel radius. It is essentially short, thus it limits the beam transport in a relatively short channel. Widening the channel can extend the length but the accelerating gradient reduces more significantly as it is roughly inversely proportional to the square of the channel radius.

The experiment at FACET [6] measured the induced transverse wakefield by a misaligned positron beam with respect to the hollow channel and confirmed that the deflection of the witness beam scales linearly with the beam offset. Note that the driving and witness bunches were originally separated from one single bunch, thus they equally deviate from the channel. It was proposed to place the trailing bunch precisely at the zero transverse field to mitigate the deflection effect. But still, the issue remains and sets a tighter constraint on the alignment accuracy for a TeV-scale energy gain. Apparently, more studies are needed to understand the tolerances on beam offsets in the hollow channel and to explore the applicability of the channel structure for the future collider.

In the following, we examine the induced transverse wakefield in a different and nonlinear beam-plasma interaction regime. As it is complicated to analyse the nonlinear wakefields analytically, we assess the resultant detrimental effects from beam offset and tilt by means of the 2D Cartesian particle-in-cell simulations.

6.3 Beam-Channel Misalignment in the Nonlinear Regime

The simulation parameters for the driving proton beam and plasma are given in Table 6.1. We only consider the case with a single and short proton driver for preliminary studies. x and y are the horizontal and vertical coordinates, respectively. $\xi = x - ct$ is the axial co-moving coordinate. With the open source full relativistic code EPOCH [7] based on a 2D Cartesian geometry, we first study the beam dynamics of an initially offset or tilted proton driver with respect to the hollow channel axis and then demonstrate how the beam-channel misalignment affects the wakefields in the accelerating region for the witness bunch. Afterwards, we propose and investigate three potential solutions especially the one employing a near-hollow plasma structure to confine the deflected witness bunch.

Table 6.1 Proton beam and plasma parameters

Parameters	Values	Units
Bunch population, $N_{\rm p}$	1.15×10^{11}	
Bunch energy, W_0	1	TeV
Energy spread, $\delta W/W$	10%	
RMS length, $\sigma_{\rm x}$	150	μm
RMS radius, $\sigma_{\rm y}$	350	μm
Angular divergence, σ_θ	3×10^{-5}	
Plasma density, $n_{\rm e}$	5×10^{14}	$\rm cm^{-3}$
Hollow channel radius, $y_{\rm c}$	350	μm

6.3.1 Beam Dynamics of the Misaligned Proton Driver

Figure 6.1 displays three cases under consideration where the proton driver initially has different states: normal (taken as the baseline, Fig. 6.1a), with a positive y offset (Δy) of 50 μm (Fig. 6.1d), with an anti-clockwise and linear tilt angle (α) of $\pi/12$ (Fig. 6.1g). In the normal case without beam-channel misalignment, the proton bunch is free from the ion defocusing within the hollow channel and strongly confined by plasma electrons concentrated at the channel boundaries (Fig. 6.1b, c), which form a reflecting-wall like focusing structure (i.e., a deep potential well as shown in Fig. 6.2b).

The misalignment of the proton driver results in asymmetric plasma electron density perturbation along the axis. For a positive beam-channel offset of 50 μm, plasma electrons concentrate more at the upper sheath of the first bubble (see the cyan areas out of the first bubble in Fig. 6.1e), where there is a higher peak density in comparison with the baseline case (Fig. 6.2a). The peak density at the other side of the bubble is obviously smaller. The uneven bubble density distribution leads to asymmetry of the transverse plasma wakefields (Fig. 6.1f), which will apparently alter the transverse evolution of the proton bunch. From Figs. 6.2b, c, we see the bottom of the potential well at the proton beam is slightly inclined pointing to the same direction of the beam offset. It suggests that the protons within the channel tend to accumulate to the upper side of the bubble and thus the beam centroid gradually moves along the positive y direction. In addition, the potential well is getting shallow with the propagation distance (Figs. 6.2c), which is the same as the baseline case. This is because the driver hasn't reached the equilibrium state in the short simulation length.

In the tilted driver case, we deliberately pick up an anti-clockwise tilt to create a plasma density perturbation in an opposite way (Fig. 6.2a). More specifically, the plasma density has a larger peak density at the down sheath of the bubble (Fig. 6.1h) and the potential well bottom is oblique towards the negative y direction (Fig. 6.2b, d). In comparison with the beam offset case, the beam tilt changes the wake potential more significantly.

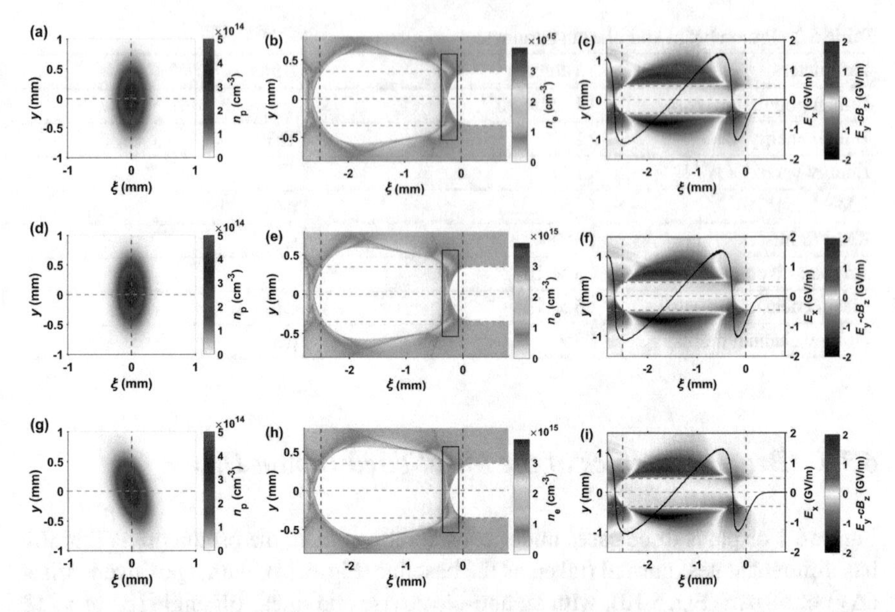

Fig. 6.1 Initial proton driver density profiles (1st column, $x = 0$), plasma electron density distribution (2nd column) and transverse plasma wakefield distribution (3rd column) along with the on-axis longitudinal electric field (black curve) when the proton bunch travels for 1.0 m (that is, $x = 1.0$ m) for three considered cases. The red and blue vertical dashed lines indicate the longitudinal centres of the driver and the witness electron bunch, respectively. The green horizontal dashed line along $y = 0$ marks the longitudinal axis while the other two off-axis ones mark the hollow channel boundaries. Note the differences between the plasma electron density distribution marked in the square frames in **b**, **e** and **h**

Based on the above discussion, we can conclude that without the defocusing plasma ions within the hollow channel, the proton driver is strongly confined within a deep potential well. Therefore, it is capable of sustaining a relatively large offset (100 μm) or tilt ($\pi/6$) without drastic distortion of the beam shape or causing beam hosing. Here we only simulate the cases for 1.0 m limited by the computing resources. Simulations for a much longer distance are important to determine whether the bunch centroid deviates observably after a long-term particle accumulation towards one direction. In that case, the bunch might get out of the potential well.

6.3.2 Wakefield Characteristics in the Witness Region

In this section, we focus on the wakefield characteristics in the second bubble where the witness electron bunch can be accelerated. Note that the electron bunch is not introduced in all simulations, as there is no quadrupole module adopted in the simulation configuration. The electron bunch if injected will radially diverge even in the

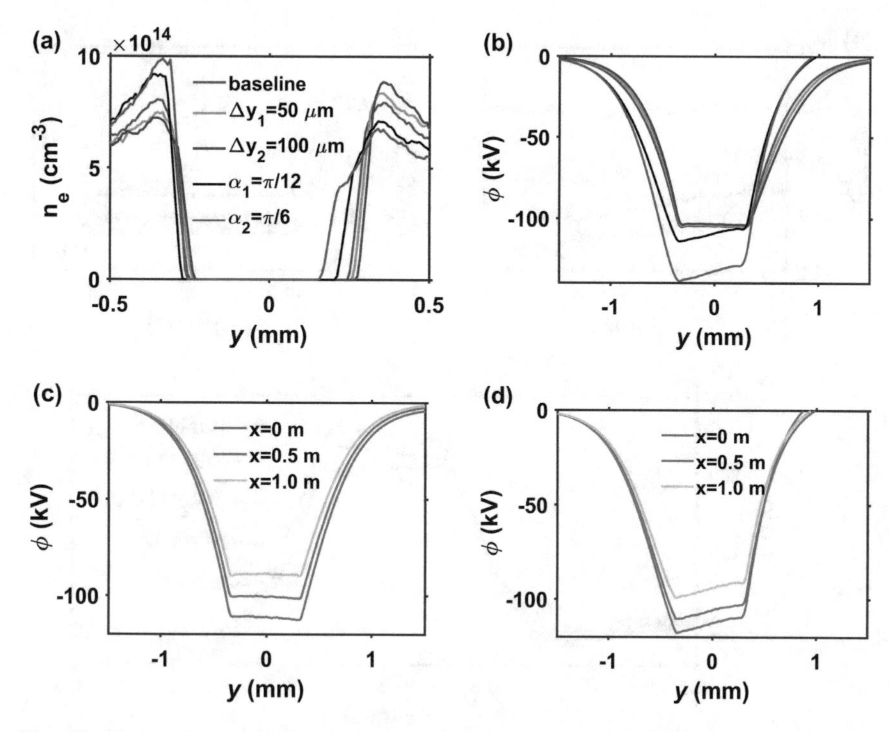

Fig. 6.2 Plasma electron densities (a) and wake potentials (b) measured along the longitudinal centre of the driver (the red vertical dashed line in Fig. 6.1) at the propagation distance of 0.5 m for different cases. The same coloured lines in **a** and **b** denote the same case. Here we include two more cases to enhance the contrast: one is with a larger beam offset ($\Delta y_2 = 100\,\mu\text{m}$) and the other is with a larger tilt angle ($\alpha_2 = \pi/6$). Wake potentials measured along the longitudinal centre of the driver at different distances are shown for the case $\Delta y_1 = 50\,\mu\text{m}$ (c) and $\alpha_1 = \pi/12$ (d)

baseline case as there is no transverse plasma focusing [1, 2]. We assume that the electron bunch misaligned or not has insignificant effect on the plasma bubble, thus the beam dynamics of the electron beam can be estimated based on the wakefield characteristics.

Figure 6.3a indicates that the driven accelerating gradient at the witness bunch decreases considerably with the beam tilt as which equivalently extends the beam length (Fig. 6.1g). The plots in the second column of Fig. 6.1 show that the offset or tilt of the driving bunch causes less asymmetry of the second bubble as it is located further away from the driver in comparison with the first bubble. Due to uneven plasma electron density distribution, the transverse wakefield in the vicinity of a potential witness bunch becomes negative (for the beam offset case) or positive (for the beam tilt case) instead of zero along the radius (Fig. 6.3b). Neglecting the calculation noises, the transverse field is uniform along the radius, which is similar to that in the linear regime [4]. The non-zero transverse field therefore will deflect the

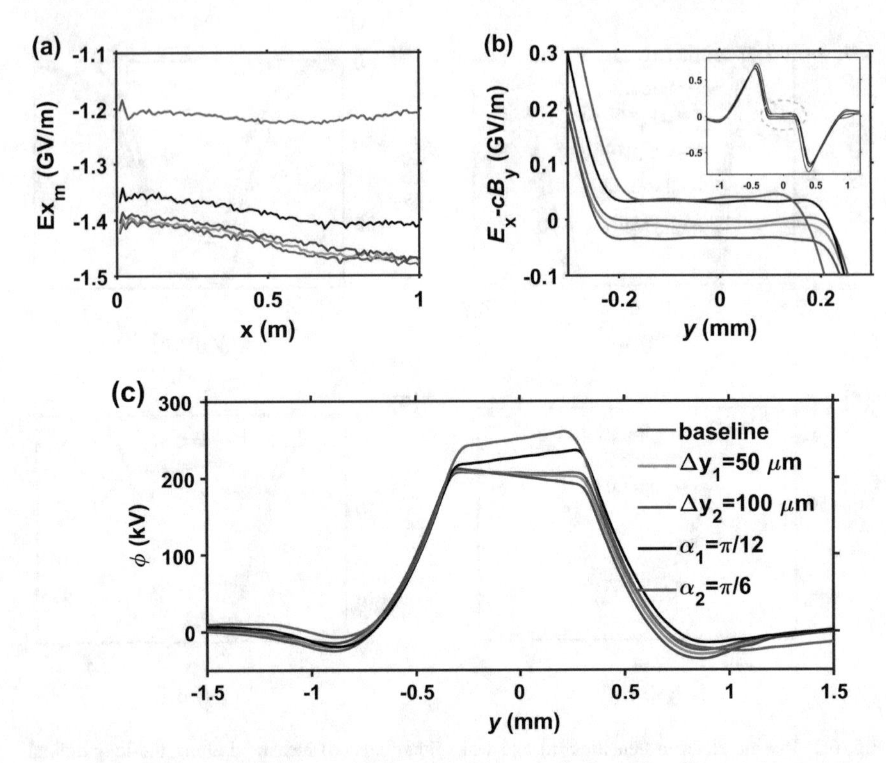

Fig. 6.3 Maximum accelerating gradient versus the proton propagation distance (a), transverse plasma wakefields along the radius (b) and wake potentials (c) measured at the longitudinal centre of the witness bunch for different driver cases at the propagation distance of 1.0 m. The inset in **b** displays the plot in a large range. The same coloured lines denote the same case

witness electrons if located. Take the case of beam offset $\Delta y_1 = 50\,\mu\text{m}$ for instance, the transverse plasma wakefield is around 15 MV/m.

Figure 6.3c gives the wake potential along the radius at the witness bunch where the channel density distribution possibly varied by the witness bunch is assumed negligible. When the driver is perfectly aligned on axis, the witness bunch see a flat potential bottom. This is exactly consistent with zero transverse plasma field. In practice, the witness bunch is stably focused by the quadrupoles fields. Nonetheless, with beam offset or tilt, the witness bunch is located in a region where the potential is radially oblique. As a result, the witness bunch will be pushed by the transverse plasma wakefields towards the positive (negative) y direction until it gets out of the accelerating bubble for the driver offset (tilt) case.

6.3.3 Mitigation of the Transverse Deflection

Based on the results above, we can conclude that the proton driver is robust to the misalignment due to the strong focusing formed at the channel wall, while the witness bunch is easily deflected by the transverse field induced as it initially has no plasma focusing. In this section we propose three solutions to potentially confine the witness bunch.

The first solution is to place the witness bunch closer to the driver so that it resides in the accelerating region where there is strong focusing at the hollow boundaries (the red regions at $\xi \in [-2.5\,\text{mm} - 1.5\,\text{mm}]$ in Fig. 6.1c, f, i). In so doing, the witness bunch can be strongly confined within the wide channel regardless of the initial driver offset or tilt. The down side is the beam emittance might be significantly disrupted while being continuously reflected by the channel boundaries.

The second solution is to add enough quadrupole focusing to the witness bunch. The issue is, the deflecting wakefield that needs to be compensated is much stronger than the focusing that the state-of-the-art quadrupoles can provide, which operates at a magnetic field gradient of about 100 T/m. For example, it requires a magnetic gradient of 1×10^3 T/m to compensate a deflecting wakefield of 15 MV/m at a radius of 50 μm (normally the size of a witness electron beam).

Given the limitations of the first two methods, we further propose to introduce a near-hollow plasma structure to replace the hollow channel. The concept of near-hollow plasma was initially proposed to separate the focusing and accelerating of the accelerated bunch and alleviate its emittance growth due to Coulomb scattering [8]. It assumes a lower density (n_{in}) plasma within the channel surrounded by a high density (n_{e}) plasma in the annulus. In general, while plasma ions located within the channel provide focusing for the witness bunch, a low witness beam emittance is still obtainable as the transverse plasma focusing is weak. In theory, the focusing electric field generated by the uniform plasma ions follows the equation

$$E_y = \frac{n_0 e}{2\epsilon_0} y, \tag{6.1}$$

where n_0 is the ion density within the channel and ϵ_0 is the vacuum permittivity. Assuming the witness bunch radius is 50 μm, to get an ion focusing which can compensate the deflecting radial field (15 MV/m) at the witness bunch caused by a driver offset of 50 μm, the plasma density n_{in} within the channel is around $3 \times 10^{13}\,\text{cm}^{-3}$, which is approximately $0.1 n_{\text{e}}$.

Figure 6.4a demonstrates that by employing a near-hollow plasma where $n_{\text{in}} = 0.1 n_{\text{e}}$, a local potential well at the witness bunch is created. The well depth is 2.5 kV from the lower well edge. As a result, the witness particle can be confined under the driver misalignment, provided that its transverse kinetic energy is smaller than 2.5 keV. It corresponds to an angular spread of 7×10^{-4} for a 10 GeV bunch, which allows for a large margin of the beam emittance for the witness bunch. More importantly, the potential well gets increasingly deep with the propagation distance, indicating an increasingly strong focusing for the witness bunch. Note that, while the

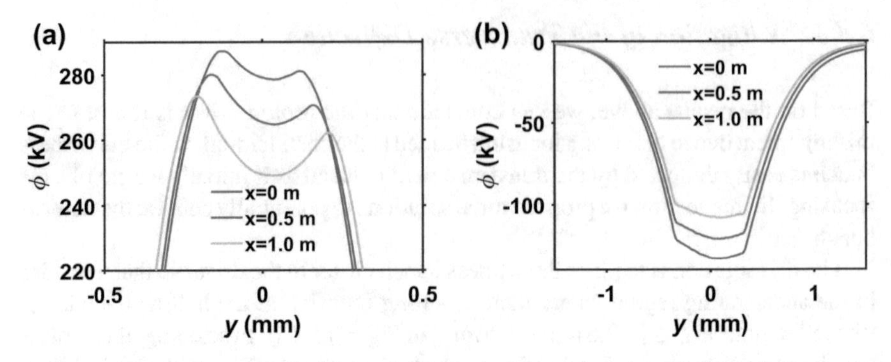

Fig. 6.4 Wake potentials at the longitudinal centers of the witness bunch (a) and the proton bunch (b) for different propagation distances for the case $\Delta y_1 = 50\,\mu\text{m}$, $n_{\text{in}} = 0.1n_{\text{e}}$

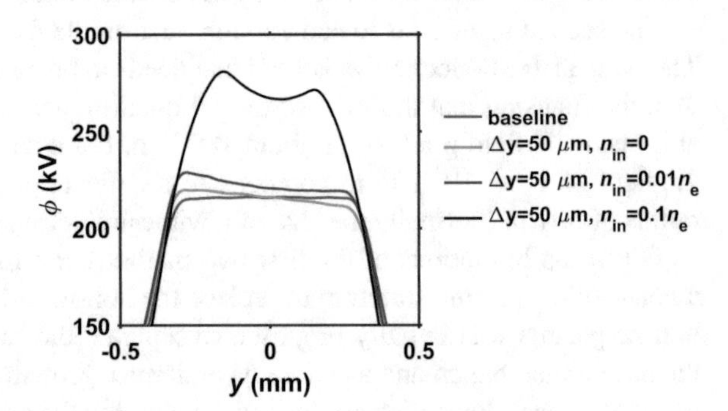

Fig. 6.5 Wake potentials at the longitudinal center of the witness bunch for different cases when the driver travels for 0.5 m

plasma ions within the channel provide focusing for the witness electrons, they are defocusing for the proton driver. Regardless, Fig. 6.4b shows a even deeper potential well, indicating strong proton focusing. This is due to more contribution from the additionally introduced plasma electrons by the near-hollow plasma. In addition, the lowest potential at the proton bunch resides almost on the axis, which mitigates the proton accumulation towards a specific direction.

Figure 6.5 includes the case when $n_{\text{in}} = 0.01n_{\text{e}}$ for comparison. We see that the resulting wake potential is similarly oblique as the beam offset case in purely hollow plasma. The reason is, apart from plasma ions, the near-hollow plasma also introduces extra plasma electrons initially located within the channel. These plasma electrons participate in the wake excitation as well, so the perturbed plasma density difference between two sides of the bubble brings additional defocusing wakefields to the witness bunch, which counteracts the focusing from the plasma ions within the channel.

6.4 Summary

In this chapter, we have demonstrated some preliminary simulation results regarding the initially offset or tilted proton bunch traveling in the hollow plasma. The proton bunch itself is less sensitive to the initial misalignment as it has strong focusing from the hollow channel, while the induced asymmetric transverse fields deflect the witness bunch and result in particle loss. Potential solutions are varying the loading position of the witness bunch, adding strong quadrupoles, and adopting a near-hollow plasma, which is the most effective one and is also promising to conserve the witness quality. Nevertheless, the tolerance on the beam-channel misalignment in the hollow channel for a long distance is not clear yet. A quasi-static PIC code based on a 2D plane geometry or a 3D geometry is feasible to simulate the long term cases to get more understandings and to further determine whether the hollow structure is suitable for future applications.

References

1. Li Y, Xia G, Lotov KV, Sosedkin AP, Hanahoe K, Mete-Apsimon O (2017) High-quality electron beam generation in a proton-driven hollow plasma wakefield accelerator. Phys Rev Accel Beams 20(10):101301
2. Li Y, Xia G, Lotov KV, Sosedkin AP, Hanahoe K, Mete-Apsimon O (2017) Multi-proton bunch driven hollow plasma wakefield acceleration in the nonlinear regime. Phys Plasmas 24(10):103114
3. Li Y, Xia G, Lotov K, Sosedkin A, Zhao Y (2019) High-quality positrons from a multi-proton bunch driven hollow plasma wakefield accelerator. Plas Phys Control Fusion 61:025012
4. Gessner SJ (2016) Demonstration of the a hollow channel plasma wakefield accelerator. PhD thesis, Stanford University, Palo Alto, California, USA
5. Chao AW (1993) Physics of collective beam instabilities in high energy accelerators. Wiley, New York
6. Lindstrfim C, Adli E, Allen J, An W, Beekman C, Clarke C, Clayton C, Corde S, Doche A, Frederico J et al (2018) Measurement of transverse wake-fields induced by a misaligned positron bunch in a hollow channel plasma accelerator. Phys Rev Lett 120(12):124802
7. [Online]. Available: http://www.ccpp.ac.uk/codes.html
8. Schroeder CB, Esarey E, Benedetti C, Leemans W (2013) Control of focusing forces and emittances in plasma-based accelerators using near hollow plasma channels. Phys Plasmas 20(8):080701

Chapter 7
Self-modulated Long Proton Bunch Driven Plasma Wakefield Acceleration

7.1 Introduction

In previous chapters, we either study the short proton bunch or the equidistant short proton bunch trains which are assumed to be obtained from longitudinal modulation of a long proton bunch. In this chapter, we focus on the long proton bunch and the seeded self-modulation which transforms the long bunch into a train of micro-bunches and enables the resonant wakefield excitation. We introduce the collaborative project—AWAKE experiment, which is based on the concept of seeded self-modulation. Given the significant decrease of the phase velocity of the self-modulated wakefields and proton loss, we propose to use a specially tapered plasma profile to promote the wakefield amplitude and validate its effect in simulations.

7.2 Physics in Self-modulated Plasma Wakefield Acceleration

7.2.1 Concept of Seeded Self-modulation

The concept of proton driven plasma wakefield acceleration is compelling due to the ability of plasmas to sustain strong accelerating fields and the ability of proton bunches to power these fields over long distances. Successful exploitation of this concept assumes the proton bunch to be as short as hundreds of μm [1–3], while currently producible ones are dozens of cm long. This is orders of magnitude longer than the plasma wavelengths of interest, which makes strong wake excitation virtually impossible. Take the SPS bunch ($\sigma_z = 12$ cm, $\sigma_r = 200$ μm) for example, to satisfy $k_p\sigma_z \approx \sqrt{2}$, the plasma density n_p is as low as 4×10^9 cm^{-3} and the corresponding wave-breaking field is only several MV/m, which is trivial for wakefield acceleration.

Y. Li, *Studies of Proton Driven Plasma Wakefield Acceleration*, Springer Theses,
https://doi.org/10.1007/978-3-030-50116-7_7

Given the challenges in compressing the proton bunches in a large scale, scientists turn to a new way to deal with this issue. In 2010, Kumar et al. [4] proposed the seeded self-modulation (SSM) of the long proton beam, whose concept is very much like the self-modulated LWFA [5, 6]. To be specific, when a long proton bunch propagates in the plasma, it generates a weak plasma wave within its body. The transverse wave field ripples the bunch itself, which further promotes the wave amplitude. With the positive loop, the periodically focusing and defocusing forces gradually modulate the long bunch into many equidistant micro-bunches which are one plasma wavelength apart. Strong wakefields are excited accordingly.

In principle, the axisymmetric self-modulation instability (SMI) starts from the noises, i.e., a low amplitude plasma wave, which is calculated to be in the order of 10 kV/m [7]. Likewise, the competing non-axisymmetric instabilities like the hosing instability also arises from noises and is very likely to destroy the beam and the plasma wave. Fortunately, with proper seeding (that is, by creating an initial wakefield with the amplitude larger than that of the undesired noises), the SMI mode grows quickly and suppresses the destructible modes significantly [8–10]. As a result, the SSM enables the micro-bunching and applicability of the long proton bunch in the PWFA scope. The seeding options include introducing a short electron bunch [7] or a short laser pulse [11] preceding the proton bunch, creating a sharp cut in the beam current profile [12], and introducing an ionization co-propagating laser in the middle of the proton bunch, which is followed by the AWAKE (Advanced Wakefield) experiment demonstrated later. Several related experiments with long or pre-bunched electron beams [12–18] are proposed or conducted to enhance the understandings of the SMI and resonantly driven waves.

As the long proton bunch sees many periods of longitudinal accelerating and decelerating wakefields during the development of SSM, it is natural to suspect if the longitudinal motion of the protons dominates the beam modulation. We can estimate the longitudinal dynamics of protons in the bunch as the proton velocity spread multiplied by the propagation time, that is,

$$\Delta z = \frac{L}{c}\Delta v = L\Delta\beta = L\frac{\Delta\gamma}{\gamma^3}, \tag{7.1}$$

where L is the propagation distance, v is the proton velocity and γ is the relativistic factor. Take the AWAKE parameters for example, the energy spread in the proton beam increases to roughly 1.25% (5 GeV out of 400 GeV) from initial 0.35% after traveling through 10 m plasma, assuming the field gradient is $\sim$0.5 GV/m. The proton longitudinal motion is thus around 0.8 μm, which is apparently negligible. Δz comparable to the plasma skin depth of 0.2 mm only happens when the propagation distance is around 160 m. However, the electron acceleration stops much earlier, because the acceleration distance (67 m calculated based on Sect. 3.2.2) is much shorter due to electron dephasing with respect to the wakefield. As a consequence, the micro-bunching in our studied scope is a transverse process. It is the transverse forces of the wakefield do the micro-bunching rather than the energy modulation coupled with the longitudinal motion.

7.2.2 *Phase Velocity of the Self-modulated Wakefield*

Since the self-modulation instability occurs and grows through the coupling between the transverse wakefields and the radial envelope of the long beam, the properties of the self-modulated wakefields are largely affected by the beam-plasma dynamics.

In the linear regime, within the framework of the beam envelope evolution, the growth rate of the SSM can be described as [19, 20]

$$\Gamma = \frac{3\sqrt{3}}{4}\omega_{\mathrm{p}}\left(\frac{n_{\mathrm{b}}m_{\mathrm{e}}}{2n_{\mathrm{e}}m_{\mathrm{b}}\gamma_{\mathrm{b}}}\frac{\xi}{ct}\right)^{1/3}, \tag{7.2}$$

where $\omega_{\mathrm{p}} = \sqrt{4\pi n_{\mathrm{e}}e^2/m_{\mathrm{e}}}$ is the overdense plasma frequency, the quantities with the subscript 'b' denote the variables for the proton beam, and the co-moving position variable is $\xi = v_{\mathrm{b}}t - z$ where $v_{\mathrm{b}} = c\sqrt{1-\gamma_{\mathrm{b}}^{-2}}$. The growth rate scales as $\Gamma \propto n_{\mathrm{e}}^{1/6} \propto Q^{1/3}$, where Q is the proton beam charge. As a result, the growth rate of SSM increases with the plasma density and the beam population, but decreases with the beam propagation (i.e., with increasing ct). In addition, the SSM grows from the bunch head to the tail (i.e., with increasing ξ).

The phase velocity of the wakefield is

$$v_{\mathrm{ph}} = v_{\mathrm{b}}\left[1 - \frac{1}{2}\left(\frac{n_{\mathrm{b}}m_{\mathrm{e}}}{2n_{\mathrm{e}}m_{\mathrm{b}}\gamma_{\mathrm{b}}}\frac{\xi}{ct}\right)^{1/3}\right], \tag{7.3}$$

which is closely connected to the growth rate of the self-modulation instability. We see the wake phase velocity can be significantly smaller than the beam velocity during the development of SSM and it is expected to be approaching the beam velocity when the instability saturates ($\Gamma = 0$). Reference [20] simulated the case with a 450 GeV SPS bunch and the relativistic factor of the lowest wake phase velocity is $\gamma_{\mathrm{ph}} = (1 - v_{\mathrm{ph}}^2/c^2)^{-1/2} \approx 38$, which is one order of magnitude smaller than the γ-factor of the drive proton bunch.

The large wake phase slowdown or backward phase shift forces a large number of protons from the focusing regions into the defocusing regions, which results in dramatic proton loss and wake amplitude decrease after reaching the peak [21–23]. In addition, it is deleterious to the trapping and acceleration of witness electrons. To be specific, the beam energy gain is subject to the dephasing length L_{d}, which corresponds to a phase shift of half plasma wavelength. Assuming the accelerated electron has a velocity close to c, it gives $(1 - v_{\mathrm{ph}}/c)L_{\mathrm{d}} = \lambda_{\mathrm{p}}/2$, i.e., $L_{\mathrm{d}} \simeq \gamma_{\mathrm{ph}}^2\lambda_{\mathrm{p}}$. Hence the energy gain is $\Delta W \simeq eE_{\mathrm{m}}L_{\mathrm{d}} \simeq 2\pi\gamma_{\mathrm{ph}}^2(E_{\mathrm{m}}/E_{\mathrm{wb}})m_{\mathrm{e}}c^2$, where E_{m} is the maximum accelerating field seen by the electron bunch and E_{wb} is the wave-breaking field. Since $E_{\mathrm{m}} \ll E_{\mathrm{wb}}$ and γ_{ph} is low, the obtainable energy gain is very limited. Another adverse effect of the wake phase shift is the electrons get scattered when they slip from a focusing phase to a defocusing one.

The issue on the witness electron bunch can be avoided by side injection [20, 24, 25]. In that scenario, low energy electrons are side-injected with a small angle and they reach the driver axis slightly before the instability saturates. Then they are sucked in the wake by the transverse fields and scheduled into the accelerating and focusing region. Another approach is to use two plasma stages so that the long proton bunch gets self-modulated until the instability saturates in the first stage and then the electron bunch is injected into the second stage for acceleration. This concept has been employed in the design for the AWAKE Run 2.

7.3 AWAKE Experiment at CERN

AWAKE experiment [26–32] is a proof-of-principle experiment dedicated to studying the plasma wakefield excitation and electron acceleration driven by the long proton beam, which is provided by the SPS. It is the world's first proton driven PWFA experiment and currently being conducted at CERN. The potential of proton driven PWFA was initially discussed in 2009 at CERN and the first AWAKE collaboration was officially formed in 2012. AWAKE is under actively collaborative study by initially over 50 scientists from 13 institutes. Now the collaboration has expanded to over a hundred scientists from 22 institutes and it keeps attracting more expertise contributed to it.

The AWAKE experiment was approved in 2013. The schematic layout of the first run is shown in Fig. 7.1, with the proton bunch from the SPS as the driver and the electron bunch from the RF gun to be accelerated. A laser pulse used to ionizing the rubidium vapour into plasma co-propagates with the long proton bunch, starting in its middle. In this scenario, only the rear half of the bunch with a sharp-cutting front edge propagates in the plasma, hence the self-modulation is strongly seeded to ensure a stable micro-bunching (see Fig. 7.2). The goal for Run 1 is to demonstrate a GeV scale electron acceleration.

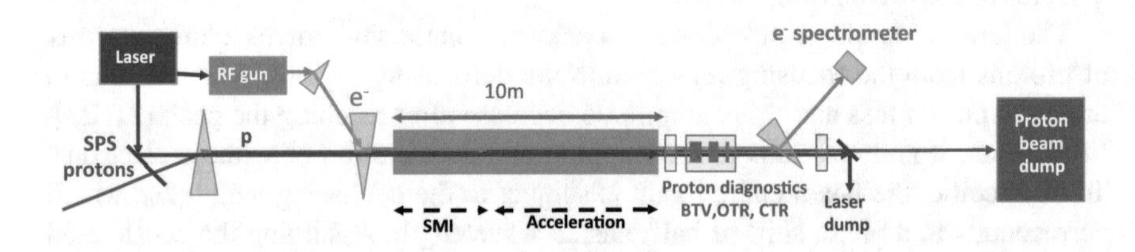

Fig. 7.1 Conceptual layout of the AWAKE Run 1 facility [31]

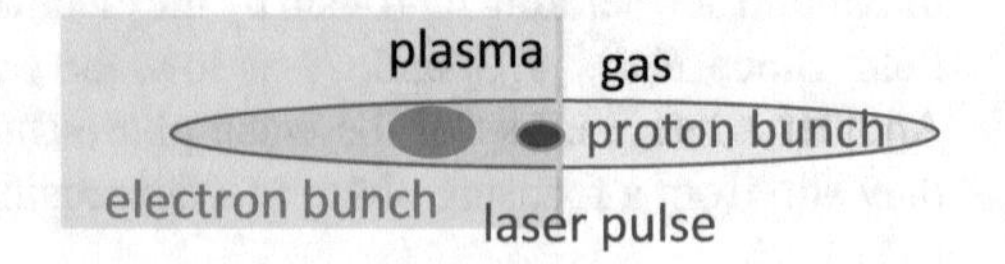

Fig. 7.2 Schematic of the self-modulation seeded by the laser ionization front

The planning for the AWAKE experiment is shown in Fig. 7.3. It starts from the baseline studies (with parameters given in Table 7.1), design, civil engineering, fabrication and installation all through the years till 2016. After commissioning of the proton beam and the laser, the AWAKE Run 1-Phase 1 (2016–2017) was conducted, which was dedicated to studying the physics of seeded self-modulation of the long proton beam. The experimental results [33, 34] are consistent with the theory. Specifically, the defocusing of protons measured by two distinct beam screens (BTV) confirm that the transverse wakefields are responsible for the self-modulation.

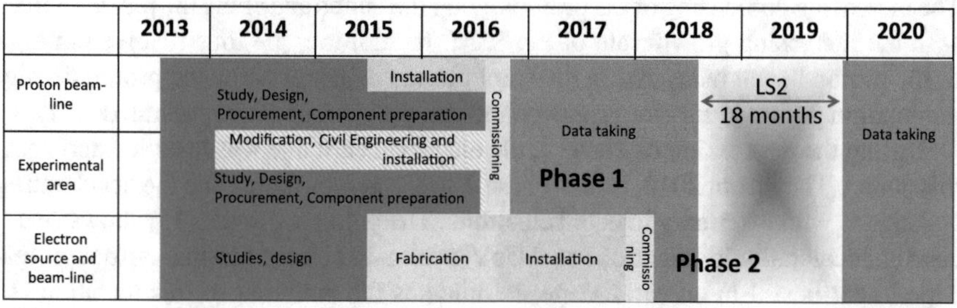

	2013	2014	2015	2016	2017	2018	2019	2020
Proton beam-line		Study, Design, Procurement, Component preparation		Installation; Commissioning	Data taking		LS2 18 months	Data taking
Experimental area		Modification, Civil Engineering and installation; Study, Design, Procurement, Component preparation		Commissioning	Phase 1			
Electron source and beam-line		Studies, design	Fabrication		Installation	Commissioning	Phase 2	

Fig. 7.3 Overview of the AWAKE experiment schedule [27]

Table 7.1 Baseline parameters of the AWAKE experiment

Parameters	Values	Units
Plasma		
Density, n_e	7×10^{14}	cm^{-3}
Ion-to-electron mass ratio (rubidium), m_i	157000	
Proton bunch		
Population, N_b	3×10^{11}	
Energy, W_b	400	GeV
Energy spread, $\delta W_b / W_b$	0.35%	
Bunch length, σ_z	12	cm
Bunch radius, σ_r	200	μm
Normalized emittance, ε_n	3.5	mm mrad
Electron bunch		
Population, N_e	1.25×10^9	
Energy, W_e	16	MeV
Normalized emittance, ε_{en}	2	mm mrad
Bunch length, σ_{ze}	0.25	cm
Bunch radius at injection point, σ_{re}	200	μm
Injection angle for electron beam, ϕ	9	mrad
Injection delay relative to the laser pulse, ξ_0	13.6	cm
Intersection of beam trajectories, z_0	3.9	m

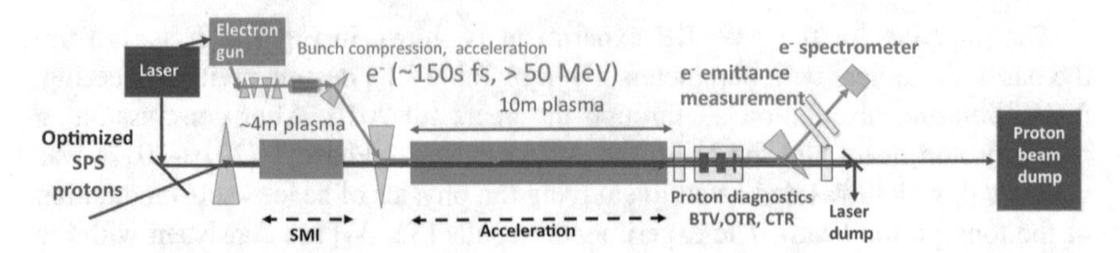

Fig. 7.4 Conceptual layout of the AWAKE Run 2 facility [32]

The increasing defocusing of the protons along the bunch and along the plasma agrees with the increasing growth rate of the SSM. In addition, the time resolved images of the proton bunch measured in different plasma densities show the proton density modulation period in inverse proportion to the square root of the plasma density.

During the data taking of Phase 1, the electron beam line was installed and commissioned. Then from 2018, Run 1 Phase 2 was conducted to probe the accelerating wakefields with externally injected electrons. The published work [35] shows a successful acceleration of electrons to 2 GeV over a 10 m long plasma. Although this experiment is still in an early stage, it validates the potential of this technique to generate very high energy electrons in a single plasma stage, which is an important step towards the development of future energy frontier particle accelerators.

Because the seeded self-modulation develops in the first several meters of plasma, it has been suggested to use two separate plasma cells (Fig. 7.4) to optimize the acceleration process. The first plasma cell is used exclusively for the beam self-modulation until it saturates and establishment of the wakefields. Then a shorter and higher energy electron beam in comparison with that in Run 1 is injected into the second cell for acceleration. Since the wake phase becomes stable in the second cell, it is easier to precisely control the electron beam within an accelerating and focusing phase. This proposal has been submitted as AWAKE Run 2 which aims at obtaining tens of GeV electron bunch with high beam quality driven by a compressed SPS bunch. It will take place after the long shut-down of the LHC.

Future experiments will include ultra-short electron bunches to cope with accelerator related issues. In a longer term, the realization of short proton bunches will enable the wakefields driven free from SMI and make it promising to accelerate electrons to the energy frontier.

7.4 Amplitude Enhancement of the Self-modulated Plasma Wakefields

As the wake phase velocity decreases significantly before the self-modulation saturates, it has been proposed to alter the longitudinal plasma profile to control the development of SSM [21–23, 36, 37]. That is, a plasma density increase is introduced to reduce the local plasma wavelength so that the backward phase shift can be

Table 7.2 Proton and plasma parameters

Parameters	Values	Units
Bunch population, N_p	3×10^{11}	
Bunch energy, W_0	400	GeV
Energy spread, $\delta W/W$	0.7%	
RMS length, σ_z	6	cm
RMS radius, σ_r	200	μm
Normalized emittance, ϵ_n	3.5	μm
Plasma density, n_0	7×10^{14}	cm^{-3}

compensated. The plasma wakefield amplitude has been promoted, but suboptimally [23]. In this section, we accommodate the need for more complicated plasma density profiles and propose a new plasma shape which can further boost the wakefield.

It is worth pointing out that the plasma tapers are different from the plasma density fluctuations [38]. Such plasma inhomogeneity leads to the phase difference between the long modulated proton driver or the witness bunch and the plasma wave. It could destroy the accelerated bunch in the defocusing region. But it has been proved to have little effect on SSM.

7.4.1 Baseline

With LCODE based on a 2D cylindrical geometry, we first simulate a twice shorter proton beam as compared to the baseline AWAKE scenario [31]. This beam length is envisaged for AWAKE upgrades (Table 7.2). The beam propagates in z-direction (to the right in Fig. 7.5a). Only the rear half that propagates in plasma is simulated, starting with an abrupt cut, so as to mimic SSM seeding by the ionization front. With periodic focusing and defocusing transverse forces excited (i.e., negative and positive wakefield potentials shown by green lines), some protons are confined (marked as red points) and others diverge (marked black). In this way, the long bunch is chopped into many micro-bunches.

Ideally, the micro-bunches are approximately $\lambda_p = 2\pi c/\omega_p$ apart, where $\omega_p = \sqrt{4\pi n_0 e^2/m_e}$ is the plasma frequency, and other notations are common. Each bunch occupies $\lambda_p/4$, corresponding to both decelerating and focusing region. However, before the micro-bunching, the wake potential near the bunch head has larger regions with negative values than with positive ones [39]. From Fig. 7.5b, c, we see protons along $\xi \in [-2\pi, 0]$ are focused and located in the negative potential region. Therefore, the first micro-bunch occupies almost a full wave period after self-modulation and shifts the wave phase back by almost $2.5c/\omega_p$ with respect to the initial perturbation (Fig. 7.5d). Note that the nonlinear elongation of the wake period due to increasing wake amplitude along the bunch is negligible here [40]. As SSM develops,

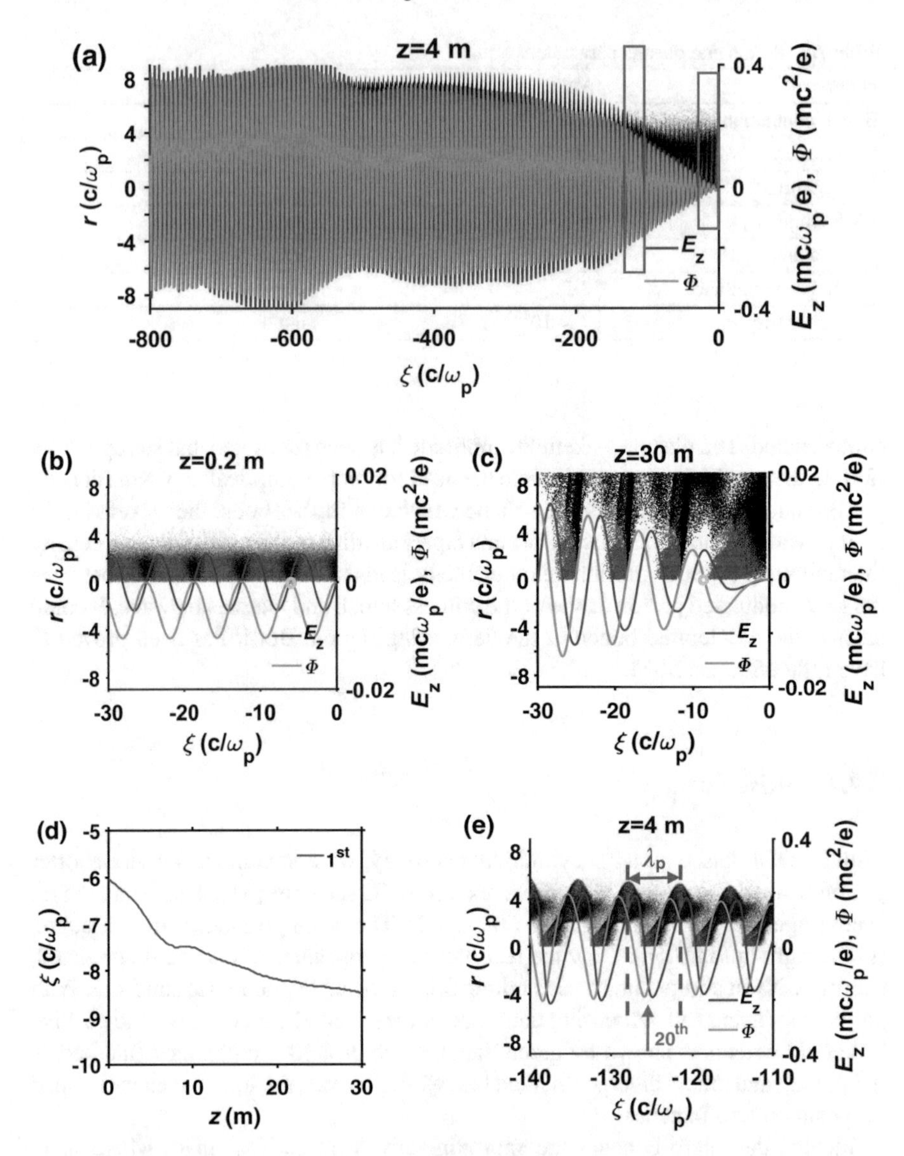

Fig. 7.5 Bunch distribution in real space (points), on-axis longitudinal electric field E_z (blue curve), and wakefield potential Φ (green curve) for the whole beam at $z = 4\,\text{m}$ (**a**), for the bunch head ($\xi \in$ [–30, 0]) at $z = 0.2\,\text{m}$ (**b**) and $z = 30\,\text{m}$ (**c**), and for the front part ($\xi \in$ [–140, –110]) at $z = 4\,\text{m}$ (**e**). Red points denote the protons that are confined by potential wells, while black points represent protons capable of escaping. **d** Longitudinal position $\xi = z - ct$ of the first potential zero [two positions therein are marked by cyan circles in **b** and **c**] versus the propagation distance z

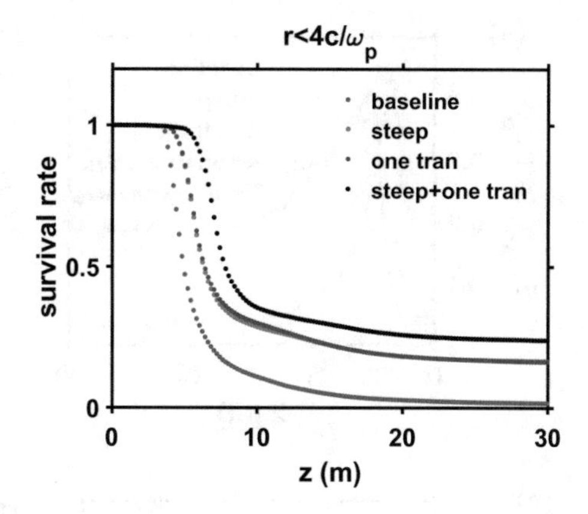

Fig. 7.6 Survival rates of protons initially located in the interval of the length λ_p near the 20th bunch for all discussed cases. A proton is counted as survived if its radial position is smaller than $4c/\omega_p$

plenty of protons drop into defocusing regions and get lost because of the significant backward phase shift. Take the 20th micro-bunch for instance (Fig. 7.5e), we see only 2% of protons survived (Fig. 7.6). Without doubt, the wakefield drops drastically with such huge proton loss (Fig. 7.7a).

7.4.2 *Plasma Profiling*

Since the drastic phase shift is the main issue hindering strong wake excitation based on SSM, researchers have proposed two typical types of longitudinal plasma density profiles: density step-up [noted as case 1, Fig. 7.8a1] and linear density transition [noted as case 2, Fig. 7.8b1]. Increase of the plasma density reduces the local plasma wavelength and thus brings forward the back shifted phase. With the same proton beam parameters given in Table 7.2 and the same initial plasma density, we first sort out the optimum plasma density profile for each case. "Optimum" here corresponds to the highest established wake amplitude that we measure at $z = 30$ m. For the case 1, the relative plasma density increases steeply by 4.5% at $z = 1.6$ m. The optimum linear transition corresponds to 6% density increase in the interval from $z = 0.2$ m to $z = 2.2$ m. Figure 7.7a confirms a substantial enhancement of the wake amplitudes by these two plasma profiles.

Here we propose a third plasma density profile. It features an early and steep density increase immediately followed by a linear transition, as shown in Fig. 7.8c1. The optimum set-up is to increase the density steeply by 1.5% after 0.2 m and then linearly until 13.5% in 2 m. It resembles a simple combination of the first two cases but is capable of further enhancing the wake amplitude by about 30% (Fig. 7.7a).

The underlying reasons for the wakefield enhancement are twofold. First of all, introducing a small and sharp increase in the early stage can correct the phase shift more effectively. Figure 7.7d demonstrates that the initial phase correction ($\Delta\xi_1$) in our proposed case (black curve) is much larger than the one ($\Delta\xi_2$) produced by the

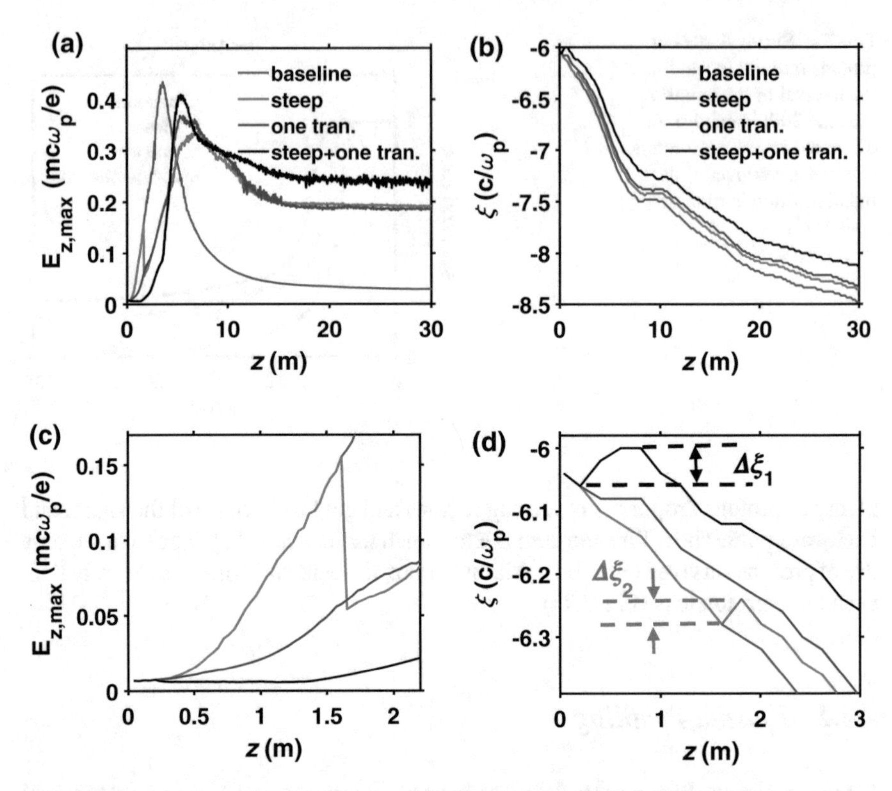

Fig. 7.7 Maximum wakefield amplitudes (**a**) and longitudinal positions of the first potential zero (**b**) versus the propagation distance for all discussed cases and their close-ups at the initial distance (**c**) and (**d**). The curves of the same colour in top and bottom plots represent the same case

later steep density increase of a larger (6%) value (case 1, green curve). Likewise, the linear density increase, although acting the same early (from $z = 0.2$ m), contributes less to the phase correction (case 2, blue curve). Another reason is, in our proposed case, the wakefield stays low during a long time (Fig. 7.7c). As a result, SSM development slows down, and protons experience smaller transverse forces in the early stage, thus gaining smaller transverse momenta [Fig. 7.8c2] than in other two cases [Fig. 7.8a2, b2]. Consequently, fewer protons have enough transverse momentum to escape from the potential well during micro-bunching. This is especially important for protons near the boundaries between positive and negative potentials, where the potential well is pretty shallow. We see a larger proton concentration near the axis and within the focusing and decelerating region [Fig. 7.8c3], which is favorable for wake excitation. As for the other two cases, protons located in the vicinity of zero potentials escape easily [marked black in Fig. 7.8a3, b3]. This is further confirmed by the micro-bunch density in Fig. 7.8c4 being observably higher than in the other two cases [Fig. 7.8a4, b4].

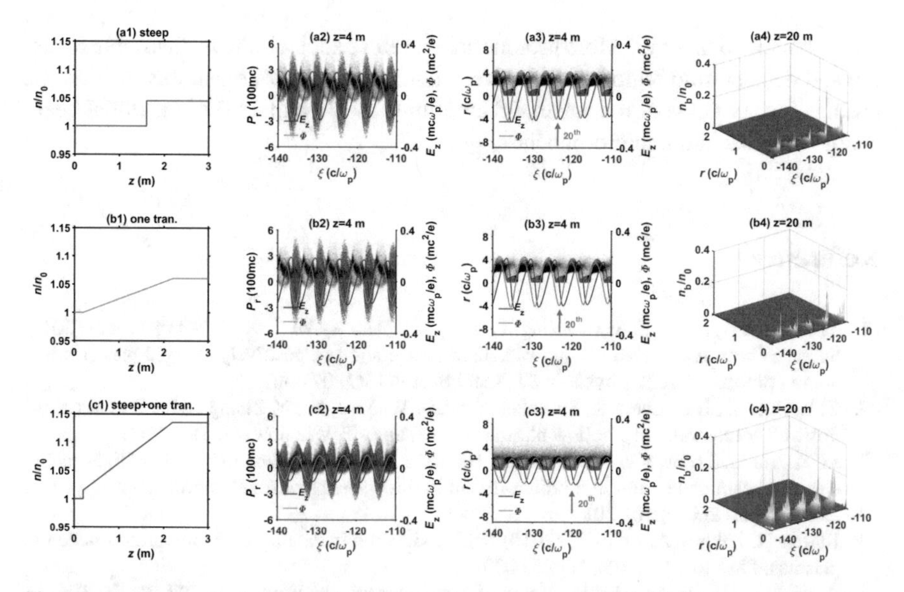

Fig. 7.8 Plasma density profiles (1st column), proton radial momenta P_r (2nd column) and positions r (3rd column) at $z = 4$ m plotted together with on-axis electric field E_z and wakefield potential Φ, and micro-bunch density $n_b(r, \xi)$ at $z = 20$ m (4th column) for three considered cases. Proton colouring is the same as in Fig. 7.5

Figure 7.6 gives a quantitative measure of the proton loss in the 20th micro-bunch for the discussed cases. In our proposed case, 24% of protons survive, which approaches the ideal value of 25% (quarter-period). For the first two cases, the survival rate is 16%. Also protons in these cases get lost quicker.

7.5 Summary

Seeded self-modulation has been demonstrated to transform a long proton bunch into many equidistant micro-bunches (e.g., the AWAKE case), which then resonantly excite strong wakefields. However, the wakefields in a uniform plasma suffer from a quick amplitude drop after reaching the peak. This is caused by a significant decrease of the wake phase velocity during self-modulation. A large number of protons slip out of focusing and decelerating regions and get lost, and thus cannot contribute to the wakefield growth. Previously suggested solutions incorporate a sharp or a linear plasma longitudinal density increase which can compensate the backward phase shift and therefore enhance the wakefields. In this chapter, we propose and investigate a slightly more complicated plasma profile. The main idea is to slow down the seeded self-modulation in the early stage, so that protons gain smaller transverse momenta while being bunched and escape less from the potential wells, especially near micro-bunch boundaries. More protons are therefore kept within the favorable focusing and decelerating region of the wave. Simulations illustrate that this new profile enables further boost of the wakefield amplitude by 30, and 24% of the initial beam charge remains in the micro-bunches.

References

1. Yi L, Shen B, Lotov K, Ji L, Zhang X, Wang W, Zhao X, Yu Y, Xu J, Wang X et al (2013) Scheme for proton-driven plasma-wakefield acceleration of positively charged particles in a hollow plasma channel. Phys Rev ST Accel Beams 16(7):071 301
2. Yi L, Shen B, Ji L, Lotov K, Sosedkin A, Wang W, Xu J, Shi Y, Zhang L, Xu Z et al (2014) Positron acceleration in a hollow plasma channel up to TeV regime. Sci Rep 4:4171
3. Li Y, Xia G, Lotov KV, Sosedkin AP, Hanahoe K, Mete-Apsimon O (2017) High-quality electron beam generation in a proton-driven hollow plasma wakefield accelerator. Phys Rev Accel Beams 20(10):101 301
4. Kumar N, Pukhov A, Lotov K (2010) Self-modulation instability of a long proton bunch in plasmas. Phys Rev Lett 104(25):255 003
5. Andreev N (1992) Resonant excitation of wakefields by a laser pulse. JETP Lett 55(10)
6. Krall J, Ting A, Esarey E, Sprangle P (1993) Enhanced acceleration in a self-modulated-laser wake-field accelerator. Phys Rev E 48(3):2157
7. Lotov K, Lotova G, Lotov V, Upadhyay A, Tückmantel T, Pukhov A, Caldwell A (2013) Natural noise and external wakefield seeding in a proton-driven plasma accelerator. Phys Rev ST Accel Beams 16(4):041 301
8. Schroeder C, Benedetti C, Esarey E, Grüner F, Leemans W (2012) Coupled beam hose and self-modulation instabilities in overdense plasma. Phys Rev E 86(2):026 402
9. Schroeder C, Benedetti C, Esarey E, Grüner F, Leemans W (2013) Coherent seeding of self-modulated plasma wakefield accelerators. Phys Plasmas 20(5):056 704
10. Vieira J, Mori W, Muggli P (2014) Hosing instability suppression in self-modulated plasma wakefields. Phys Rev Lett 112(20):205 001
11. Siemon C, Khudik V, Austin Yi S, Pukhov A, Shvets G (2013) Laser-seeded modulation instability in a proton driver plasma wakefield accelerator. Phys Plasmas 20(10):103 111
12. Vieira J, Fang Y, Mori W, Silva L, Muggli P (2012) Transverse self-modulation of ultra-relativistic lepton beams in the plasma wakefield accelerator. Phys Plasmas 19(6):063 105
13. Fang Y, Yakimenko V, Babzien M, Fedurin M, Kusche K, Malone R, Vieira J, Mori W, Muggli P (2014) Seeding of self-modulation instability of a long electron bunch in a plasma. Phys Rev Lett 112(4):045 001
14. Gross M, Brinkmann R, Good J, Grüner F, Khojoyan M, de la Ossa AM, Osterhoff J, Pathak G, Schroeder C, Stephan F (2014) Preparations for a plasma wakefield acceleration (PWA) experiment at PITZ. Nucl Instrum Methods Phys Res Sect A 740:74–80
15. Lishilin O, Gross M, Brinkmann R, Engel J, Grüner F, Koss G, Krasilnikov M, de la Ossa AM, Mehrling T, Osterhoff J et al (2016) First results of the plasma wakefield acceleration experiment at PITZ. Nucl Instrum Methods Phys Res Sect A 829:37–42
16. Adli E, Olsen VB, Lindstrøm C, Muggli P, Reimann O, Vieira J, Amorim L, Clarke C, Gessner S, Green S et al (2016) Progress of plasma wake-field self-modulation experiments at FACET. Nucl Instrum Methods Phys Res Sect A 829:334–338
17. Apsimon OM, Xia G, Burt G, Hidding B, Hanahoe K, Smith J (2016) iMPACT, undulator-based multi-bunch plasma accelerator. In: Proceedings of IPAC16, p 2609
18. Pompili R, Anania M, Bellaveglia M, Biagioni A, Bisesto F, Chiadroni E, Cianchi A, Croia M, Curcio A, Di Giovenale D et al (2016) Beam manipulation with velocity bunching for PWFA applications. Nucl Instrum Methods Phys Res Sect A 829:17–23

19. Schroeder C, Benedetti C, Esarey E, Grüner F, Leemans W (2011) Growth and phase velocity of self-modulated beam-driven plasma waves. Phys Rev Lett 107(14):145 002
20. Pukhov A, Kumar N, Tückmantel T, Upadhyay A, Lotov K, Muggli P, Khudik V, Siemon C, Shvets G (2011) Phase velocity and particle injection in a self-modulated proton-driven plasma wakefield accelerator. Phys Rev Lett 107(14):145 003
21. Lotov K (2011) Controlled self-modulation of high energy beams in a plasma. Phys Plasmas 18(2):024 501
22. Caldwell A, Lotov K (2011) Plasma wakefield acceleration with a modulated proton bunch. Phys Plasmas 18(10):103 101
23. Lotov K (2015) Physics of beam self-modulation in plasma wakefield accelerators. Phys Plasmas 22(10):103 110
24. Kalmykov SY, Gorbunov LM, Mora P, Shvets G (2006) Injection, trapping, and acceleration of electrons in a three-dimensional nonlinear laser wakefield. Phys Plasmas 13(11):113 102
25. Lotov KV (2012) Optimum angle for side injection of electrons into linear plasma wakefields. J Plasma Phys 78(4):455–459
26. Muggli P, Pukhov A, Gschwendtner E, Caldwell A, Wing M, Lotov K, Reimann O, Bracco C, Pardons A, Tarkeshian R et al (2013) Physics of the AWAKE project
27. Caldwell A, Gschwendtner E, Lotov K, Muggli P, Wing M (2013) AWAKE design report: a proton-driven plasma wakefield acceleration experiment at CERN. Technical report
28. Assmann R, Bingham R, Bohl T, Bracco C, Buttenschön B, Butterworth A, Caldwell A, Chattopadhyay S, Cipiccia S, Feldbaumer E et al (2014) Proton-driven plasma wakefield acceleration: a path to the future of high-energy particle physics. Plasma Phys Control Fusion 56(8):084 013
29. Bracco C, Gschwendtner E, Petrenko A, Timko H, Argyropoulos T, Bartosik H, Bohl T, Müller JE, Goddard B, Meddahi M et al (2014) Beam studies and experimental facility for the AWAKE experiment at CERN. Nucl Instrum Methods Phys Res Sect A 740:48–53
30. Caldwell A, Adli E, Amorim L, Apsimon R, Argyropoulos T, Assmann R, Bachmann A-M, Batsch F, Bauche J, Olsen VB et al (2016) Path to AWAKE: evolution of the concept. Nucl Instrum Methods Phys Res Sect A 829:3–16
31. Gschwendtner E, Adli E, Amorim L, Apsimon R, Assmann R, Bachmann A-M, Batsch F, Bauche J, Olsen VB, Bernardini M et al (2016) AWAKE, the advanced proton driven plasma wakefield acceleration experiment at CERN. Nucl Instrum Methods Phys Res Sect A 829:76–82
32. Adli E (2016) Towards AWAKE applications: electron beam acceleration in a proton driven plasma wake. In: Proceedings of IPAC16, (Busan, Korea) JACOW, Geneva, Switzerland, pp 2557–2560
33. Turner M et al (AWAKE Collaboration) (2019) Experimental observation of plasma wakefield growth driven by the seeded self-modulation of a proton bunch. Phys Rev Lett 122(5):054801
34. Adli E et al (AWAKE Collaboration) (2019) Experimental observation of proton bunch modulation in a plasma at varying plasma densities. Phys Rev Lett 122(5):054802
35. Adli E, Ahuja A, Apsimon O, Apsimon R, Bachmann A-M, Barrientos D, Batsch F, Bauche J, Olsen VB, Bernardini M et al (2018) Acceleration of electrons in the plasma wakefield of a proton bunch. Nature 561(7723):363
36. Schroeder C, Benedetti C, Esarey E, Grüner F, Leemans W (2012) Particle beam self-modulation instability in tapered and inhomogeneous plasma. Phys Plasmas 19(1):010 703
37. Petrenko A, Lotov K, Sosedkin A (2016) Numerical studies of electron acceleration behind self-modulating proton beam in plasma with a density gradient. Nucl Instrum Methods Phys Res Sect A 829:63–66
38. Lotov K, Pukhov A, Caldwell A (2013) Effect of plasma inhomogeneity on plasma wakefield acceleration driven by long bunches. Phys Plasmas 20(1):013 102
39. Gorn A, Tuev P, Petrenko A, Sosedkin A, Lotov K (2018) Response of narrow cylindrical plasmas to dense charged particle beams. Phys Plasmas 25(6):063 108
40. Lotov K (2013) Excitation of two-dimensional plasma wakefields by trains of equidistant particle bunches. Phys Plasmas 20(8):083 119

Chapter 8
Conclusions and Outlooks

Plasma-based wakefield accelerators have attracted enormous attention and developed significantly since the proposal in 1979, due to ultra-high accelerating gradients sustained by plasmas in comparison with those in metal cavities of radio-frequency accelerators. So far electrons have been demonstrated in experiment to be accelerated to as high as 85 GeV. Nevertheless, the achievable energy gain of the witness bunch in one acceleration stage is subject to the transformer ratio limit and thereby to the energy of the drive beam. Combining multiple acceleration stages can, in principle, overcome the transformer ratio limit, but it requires tens to hundreds stages for energy frontier collider applications, which is technically challenging. Under this circumstance, proton driven plasma wakefield acceleration looks particularly promising as existing proton bunches have huge energy and large populations, therefore they are capable of powering particles to the TeV energy level in a single and long distance plasma stage.

On the other hand, unlike the electron or laser driver, a positively charged driver attracts plasma electrons instead of fully expelling them away and forming an electron-free bubble. The resulting transverse wakefield is radially nonlinear and varies both in time and along the witness bunch. The variations of radial forces deteriorate the emittance of the accelerated beam and make it far from suitable for prctical applications in future high luminosity colliders. In addition, currently produciable high-energy proton bunches are tens of cm long, which is much longer than the interesting submillimeter plasma wavelength. Such long proton bunches therefore hardly excite strong plasma wakefields directly. Longitudinal compression of the bunch length to the plasma wavelength by traditional methods is conceivable, but it will be too costly and technically challenging to implement given the required compression factor of up to several orders of magnitude. Fortunately, the seeded self-modulation forces the micro-bunching of the long proton bunch and therefore allows harnessing it as the driver to resonantly excite strong plasma waves.

In this thesis, with theory and particle-in-cell simulations, we expand the scope of proton driven plasma wakefield acceleration from a single short proton bunch driven scheme to multiple proton bunch driven case and demonstrate the generation

Y. Li, *Studies of Proton Driven Plasma Wakefield Acceleration*, Springer Theses,
https://doi.org/10.1007/978-3-030-50116-7_8

of either a high quality TeV-scale electron or positron bunch, which is essentially favored by future high-energy lepton colliders. The hollow plasma channel involved plays a vital role in creating a beneficial accelerating structure for the acceleration and the beam quality preservation of the witness beam. Hence, it is important that we assess the deleterious deflection effects induced by the misalignment between the beam and the hollow channel. In addition, we investigate the self-modulated long proton bunch and promote the excited wakefield amplitude by a sophisticated plasma taper. Our work can be divided into five aspects:

1. Simulations of the electron acceleration driven by an assumed single short proton bunch in a hollow plasma channel demonstrate the efficient acceleration of the witness electron bunch to 0.6 TeV with a preserved normalized emittance in a single stage of 700 m. The electron beam carries the charge of about 10% of 1 TeV proton driver charge. This work is complementary to the previously published work on positron acceleration, though the discovered similarly favorable accelerating regime created in the hollow channel is not typical for nonlinear wakefields. It even contrasts with the uniform plasma case, where the accelerating structure is strongly charge dependent. In our discovered regime, the witness bunch is positioned in the region with a strong and radially uniform accelerating field, but completely free from plasma electrons and ions (i.e., no transverse plasma fields). As a result, the beam emittance is conserved and the beam energy gain depends only on the longitudinal position along the bunch, thus enabling minimization of the energy spread by tailoring the bunch shape. The quadrupoles used to confine the driver from emittance-driven widening also play an important role in witness confinement. They keep the witness bunch from entering the defocusing region that appears at large radii as the driver depletes.
2. In comparison with a single short proton bunch, which is necessarily obtained by a significant compression, it is less challenging to produce a train of short, lower charge proton bunches via longitudinal modulation of a long proton bunch. Hence, we simulate the wakefield excitation driven by multiple proton bunches. By introducing a hollow plasma channel, we remove the defocusing from the plasma ions and create a basin-like potential well by the plasma electrons attracted within the channel to confine the proton bunches in the blowout regime. Up to 90% of TeV protons from ten proton bunches survive after propagating through a plasma of length 150 m. In addition, up to half a wakefield period is applicable to each proton bunch in terms of stable deceleration and maintenance. Strong plasma waves are resonantly excited, which accelerate the witness electrons to 0.47 TeV. The hollow channel also brings beneficial features for the accelerated beam, such as emittance preservation and low energy spread (1.3%). The studied regime is attained at a proper ratio among plasma skin depth, driver radius, hollow channel radius, and micro-bunch period. It can also sustain relatively large plasma density inhomogeneity (at the level of 5%).
3. Nevertheless, the acceleration of positrons is different in the multiple proton bunch driven hollow plasma scheme because the accelerating structure is strongly charge dependent. The plasma electrons, used to provide focusing to the multiple proton

bunches, now dilute the positron bunch. It results in a trade-off between keeping a small normalized emittance and a small energy spread. By loading an extra electron bunch to repel the deleterious plasma electrons and meanwhile reducing the plasma density slightly to shift the accelerating phase with a conducive slope to the positron bunch, the positron bunch can be accelerated to 400 GeV (40% of the driver energy) with an energy spread as low as 1% and well preserved normalized emittance.

4. In terms of the non-uniformities or imperfections in the 3D world, the most severe issue is the misalignment between the beam and the plasma channel. Some preliminary simulations have been done to assess this effect induced by the initially offset or tilted driving proton bunch. The proton bunch itself is less sensitive to the initial misalignment as it has strong focusing from the hollow channel, while the induced asymmetric transverse field drastically leads to the deflection and eventually loss of the witness bunch. Adopting a near-hollow plasma is promising to confine the witness bunch, but the viability of a long term tolerance of the misalignment needs to be further explored in order to determine whether the hollow structure is suitable for future applications.
5. The amplitude of the seeded self-modulated plasma wakefield decreases greatly after reaching the peak due to a significant decrease of the wake phase velocity during self-modulation and a huge proton loss. With a sophisticated plasma taper which features an early and steep density increase immediately followed by a linear transition, we are able to boost the wakefield amplitude by 30%. In addition, 24% of the initial beam charge remains in the micro-bunches. The main idea is to slow down the seeded self-modulation in the early stage, so that protons gain smaller transverse momenta while being bunched and escape less from the potential well. More protons are therefore kept within the focusing and decelerating region, contributing to the wake excitation.

The success of the AWAKE experiment for the first time verifies the principle of plasma wakefield acceleration driven by high energy protons. It is an important step towards the future energy frontier colliders. Our work essentially expands this concept and more importantly we demonstrate the feasibility of generating both high energy and high quality electrons and positrons, which are promising candidates for the future lepton colliders. To assess our proposed schemes in experiment, it apparently needs a lot of further efforts. Specifically, it requires advanced technology development to produce one short proton bunch or a proton bunch train. The experimental implementation of a long hollow channel is challenging as the current development is still in its infancy. Another important issue that requires much attention is the adverse transverse effects induced by the beam-channel misalignment, which could destroy the whole acceleration process. We have preliminarily studied it for a short distance, whereas the long term effects need to be determined. In addition, in our simulations we assume the part within the hollow channel pure vacuum, but it is actually filled with unionized gas if the channel is created by gas ionization from a shaped laser. All these issues should be investigated further before this hollow-plasma based acceleration scheme becomes viable in the future, generating high quality beams for compact and affordable energy frontier lepton colliders.

Printed in the United States
by Baker & Taylor Publisher Services